SAFE SYSTEM
OR
STALINIST SYSTEM?

Road Safety at Any Cost

Rob Morgan
F.I.T.E., F.A.I.T.P.M., A.R.P.S.

Published by Robert Morgan
4 Kent Court
Bulleen Victoria 3105

National Library of Australia Cataloguing-in-Publication data:
Safe System or Stalinist System? Road Safety at Any Cost
ISBN: 978-0-9581139-1-5

First edition: July 2018
A conference paper with the title *Safe System or Stalinist System* was presented at the 2018 National Conference of the Australian Institute of Traffic Planning and Management (AITPM) in Perth, Western Australia, July 2018. This book is the full story.

Cover design: Luke Harris
Editor: Anthea Wynn
Typeset: WorkingType Studios
Printed in Australia by Ingram Spark

To GMLQ

I first met Geoff Quayle in 1979 when I worked at the Federal Office of Road Safety for six months, before it shifted to Canberra and I shifted to local government traffic engineering. Geoff had recently completed the Office's first occasional report OR1 on rules of precedence at intersections and he was rightly proud of it. It displayed his great intellect, knowledge and amazing memory for detail. I lapped it up. After a few years I lost contact with Geoff, only to renew it in early 2017 after I read his professional memoir *Driving Past*. You will see many references to his book in this book of mine. Geoff died unexpectedly in July 2017.

CONTENTS

INTRODUCTION

Many ideas are a product of their time.

The Safe System in road safety is one such idea. Put together from an earlier Swedish idea called Vision Zero, the Safe System is a strategic approach that originated in Victoria, Australia in 2004 and has spread across the nation and far beyond.

This didn't happen when Australia's road fatality rate was the worst in the world in 1970. It happened about 10 or 15 years after economic rationalism and neoliberalism took hold, with governments downsizing the road authorities and other agencies that had been instrumental in getting the road toll under control.

At the core of the Safe System is the notion that road safety endeavours should aim to completely eliminate fatal and serious injuries on our roads. There is a physical limit to the forces a human body can withstand. So under the Safe System those forces need to be managed to avoid serious injury or death in any road crash. Ultimately most of the effort with the Safe System is with managing vehicle speeds and most of that management is through speed limits and enforcement.

Victoria has an enviable record in road safety, being the first jurisdiction in the world to mandate the wearing of seatbelts and the first to introduce random breath testing to detect the blood alcohol concentration in drivers. Victoria has also had a fine tradition of building high-standard road infrastructure with many safety features built in. While New South Wales was building more miles of poorer-design roads, Victoria was building perhaps fewer miles of roads, but of a better standard, leaving us with a great legacy.

But Victoria has also had some great episodes of Stupid.

Historian Geoffrey Blainey is quoted by Quayle, 2015 discussing the lead up to the Eureka uprising in 1854, during the Victorian gold rush:

> Whereas the government in Sydney [NSW] handled the gold crisis with masterly commonsense, the new government in Melbourne [Victoria] continued to handle its own gold crisis with less skill.
>
> Governor LaTrobe retired to Europe and was replaced by a starchy and aggressive governor, Commodore Charles Hotham. He inherited LaTrobe's inept policies on how to cope with the invasion of diggers but did not change the policies. A man of the quarterdeck, he simply enforced them. As nearly half the [77,000] miners ... succeeded in evading the licence fee, Hotham eventually ordered the police to hunt for unlicensed diggers twice a week.
>
> The subsequent rebellion was crushed, but public opinion was not. Thirteen miners, including an American Negro, were tried for high treason. The jury refused to convict them.[1]

From the 1920s to the 1950s the feared infectious disease polio was prevalent in Australia. It was a virus that attacked nerves and muscles and led to muscle loss, paralysis or worse. The established medical treatment was to immobilise patients in splints for months to avoid deformities and then start therapeutic exercise — and also conduct corrective surgery if required. The foremost Australian advocate of this method was a Victorian, Dr Jean Macnamara. Meanwhile, in Townsville, Sister Elizabeth Kenny developed an alternative treatment involving gentle massaging and stretching of muscles from the start

---

1   Blainey G, 2006, *A History of Victoria*, Cambridge University Press, Cambridge, UK as quoted in Quayle, 2015.

of an infection, with hot packs to relieve pain, but with minimal constraint. So entrenched were the attitudes of most in the Australian medical establishment, they had no time for Kenny's methods, despite her good results. Kenny's techniques became accepted in New Zealand, the USA and other countries where polio epidemics occurred and they did much to reduce the impacts of the infection on patients. But not in Australia:

> Australia was conservative and slow to change under the influence of social evolution. Nowhere was that more apparent than in Victoria, where the BMA [British Medical Association] forbade 'consultation with irregular practitioners' and employed the same tactics as it had earlier done with the homeopaths to overwhelm and eliminate the Kenny method of treating polio from within the state, and ultimately from Australia.[2]

So when I contracted polio (thankfully a mild case) in Melbourne in 1954 I was immobilised for months, then the physiotherapists got to work on my dwindled leg and foot muscles and finally, surgeons at the Royal Children's Hospital did their best on my left foot, with results that were worse than useless. Had I been in New Zealand, chances are my treatment would have been in line with Kenny's methods, my left leg and foot would have worked better throughout my life and my left foot would have looked more normal than it does. The kindest thing I can say about the experts who determined my treatment is *They meant well.*

More recently, and more relevant to the subject of this book, the NSW Centre for Road Safety advises that in New South Wales:

> Warning signs for mobile speed camera vehicles ensure that motorists see and recognise the enforcement areas. Mobile speed camera vehicles are marked and operators place portable warning signs 50 metres before and after the vehicle. Another warning is provided up to 250 metres before the vehicle. Warning signs ensure

---

2   Highly, 2016, p 4.

that all motorists see and recognise the enforcement activity. The signs also encourage motorists to stay within posted speed limits.[3]

This is overt enforcement. Meanwhile in Victoria it was reported by Campbell in late 2016 that:

V ctorian motorists may soon be caught speeding without seeing who snapped them.

The new guidelines that came into effect overnight make it legal for speed camera operators to hide behind bushes and signs, and also to trap drivers on hi ls, where the speedometer can creep away as the car gains momentum — even when cruise control is engaged.

There is no restriction from a technical, legislative or enforcement perspective on a mobile road safety camera being operated on a slope, hill or gradient, the modified rule states.

Victoria's mobile speed camera vehicles are unmarked. There are no advance warning signs. This is covert enforcement. It invites road rage.[4]

Welcome to Victoria. Let the story begin . . .

---

3   NSW Centre for Road Safety: *http://roadsafety.transport.nsw.gov.au/ speeding/speedcameras/index.html*. Image from the same website (Jan. 2018). (Image courtesy of NSW Centre for Road Safety).

4   In the same article it was reported that in the previous 12 months there were 247 incidents where mobile speed camera operators were intimidated or threatened. Approximately 110 of those involved vehicles being swerved at them in an intimidatory fashion.

Chapter 1

MANIFESTO DESTINY

In June 2017 I bought a book written by respected Australian road safety experts Ian Johnston and Eric Howard with Carlyn Muir.[5] It is perhaps the only book that sets out in detail a current view of what needs to happen in Australia to improve road safety. By coincidence, I bought another book that month, about 19th century philosopher, Karl Marx (Stedman Jones, 2016). Worlds apart you may think, although . . .

Johnston, Muir and Howard advocate for the utopian vision of zero fatalities and serious injuries on our roads. The means they promote for achieving this is the Safe System.[6]

Karl Marx had his own idea of utopia. He was a determined man, the most prominent socialist theorist of his time. He is best remembered for his co-authorship with Friederich Engels of the Communist Manifesto in 1848. It had the vision of a workers' utopia with freedom from oppression, allowing the free development of each person. The means of achieving this was communism which required, amongst other things, the abolition of private property and the destruction of capitalism, because wealthy capitalists were the ones who oppressed the masses. It was expected that workers would rise up against the

---

5   Johnston et al, 2017. It was originally published in hardback in 2014.
6   What is the Safe System? Figure 1.1 shows the Safe System framework, while the essential elements of it are described after that diagram. The Safe System is now the cornerstone of all Australian jurisdictions' road safety strategies and has been adopted by several other countries and the OECD. As part of it, all Australian state and territory road safety strategies now include the long-term vision of zero fatalities and serious injuries.

moneyed classes to create this classless society. Engels even claimed that through Marx's work, socialism was a science.

According to Stephen Kotkin, 2017:

> [Marx] saw class struggle as the great engine of history. What he called feudalism would give way to capitalism, which would be replaced in turn by socialism and, finally, the distant utopia of communism.

We can now, 170 years later, see that Marx' perspective was far too simplistic and while it looked relevant within the European political landscape of the 19th century, it was fundamentally flawed — and by no means a science. Marx could never have imagined the 20th century oppression, starvation and mass murder that resulted from his ideas for a workers' utopia. It has been conservatively estimated that under Stalin the Soviet Union suffered 3 million deaths through imprisonment, execution or mass murder and 7 million deaths through starvation, all in the name of communism. It's a classic example of the law of unintended consequences.

Inevitably, communist regimes realised the class struggle against capitalists could only succeed if all dissent was eliminated. If there is only one view of how to reach utopia, all other views must be crushed. Uniformity of thought is required. Political opposition or dissent can have no legitimacy. In communist Russia a vast secret police apparatus was set up.

As Stephen Kotkin puts it:

> [If] we've learned one lesson from the communist century, it is this: That to implement Marxist ideals is to betray them. Marx's demand to 'abolish private property' was a clarion call to action — and an inexorable path to the creation of an unchecked state.

Let me now describe what I shall call the *Safe System Manifesto*, the collection of ideas, proposals and recommendations set out in Johnston et al's book *Eliminating Serious Injury and Death from Road Transport, A Crisis in Complacency*:

▶ The vision is the utopia of zero deaths and serious injury on our roads (obvious from the title[7])

▶ The means for achieving this is the Safe System (see their Chapter 6)

▶ Safety on our roads is everything (e.g. see p 26 of Johnston et al)

▶ Human life is sacrosanct (see p 28)

▶ There can be no compromise: zero is not negotiable (e.g. see p 37)

▶ The Safe System is the breakthrough advance in traffic safety thinking (e.g. see pp 81, 155)

▶ The core consideration of the Safe System is the limited tolerance of the human body to physical force. This requires managing crash energy (see p 81)[8]

▶ Speed management is the new component compared with earlier thinking (see p 81). Speeds are too high in Australia (e.g. see p 72).

▶ Speed limits must match the level of protection (e.g. see p 118). If they don't, speed limits must be lowered (e.g. see p 72) — or if I may re-phrase it, the Safe System requires the destruction of the old speed limit regime.

---

7   Johnston et al, 2017 are advocates for this. The origin of this vision was the Swedish Parliament's acceptance of *Vision Zero* in 1997 — about which I will discuss more later.

8   Also see OECD/ITF, 2008. At p 114 it states 'The key to this is safer speeds'.

▶ The new speed limits must be intensely enforced (e.g. see pp 139, 141)

▶ Success requires the public to be convinced, so they will demand more from politicians (e.g. see pp 3, 31, 37).

The general tone of this sounds ominously familiar. There is already a uniformity of thought on these matters: all Australian jurisdictions have adopted the Safe System and all have embraced the ultimate vision of zero deaths and serious injuries. And some of it is extreme; the New South Wales *Towards Zero* road safety strategy, for example, states 'Any death or serious injury on our roads is one too many.' This is the sort of comment we might expect to see in a letter to the editor of a daily newspaper from someone with a single-issue complaint and no responsibility to think about the cost-effective expenditure of public money. Instead, it comes from the NSW Centre for Road Safety, the state government agency with a lead role in road safety.[9]

Stephen Kotkin further states about the arrival of communist rule in Russia:

> The dispossession of capitalists also enriched a new class of state functionaries, who gained control over the country's wealth. All parties and points of view outside the official doctrine were repressed, eliminating politics as a corrective mechanism.

Since around 1990 the downsizing of Australian road authority activities and the outsourcing of basic tasks has seen experience, expertise and skills disappear. In road safety, the void created has allowed new simplistic and naïve ways of thinking to gain favour. A new class of state functionary has gained control of road safety thinking. Others simply follow. In the same way that collectivised property in Soviet Russia empowered the state and controlled the people, the collectivised 'SafeSystemThink' is leading to greater control of people by the state — with little to show by way of road safety benefits.

9   *http://roadsafety.transport.nsw.gov.au/campaigns/towards-zero/index.html* (Dec. 2017).

Will we need to wait 170 years to realise that to implement the Safe System ideals is to betray what we know is most effective in road safety? The essential elements of the Safe System, described after Figure 1.1, appear benign or benevolent enough but are actually neither. Are Australians so compliant and law-abiding that they will let the Safe System happen to them without complaint?

I will explain how Australia's Safe System is at odds with what we know succeeds in road safety. I will describe its unintended adverse consequences that will affect most of us. I will finish by proposing an alternative basis for improved road safety in Australia, one with fewer unintended consequences.

The Safe System is discussed throughout this book. Let me start by illustrating its basic framework in Figure 1.1. This is an Australian (and originally a Victorian) construct, developed in 2004.

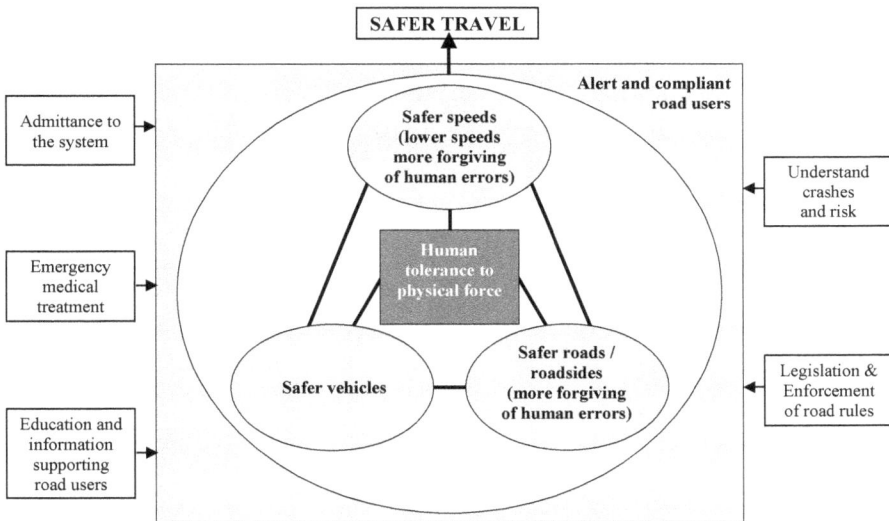

Figure 1.1 The Safe System Framework[10]

---

10   Based on Figure 6.5 in Johnston et al, 2017 where it is attributed to Eric Howard. It appears in OECD/ITF, 2008 as 'Australia's Safe System — Illustrating the primacy of speed management', without the *Emergency medical treatment* box on the left side.

From Figure 1.1 it can be seen that the essential elements of the Safe System for road safety are:

▶ At its core: human tolerance to physical force. In any crash, forces that would lead to serious injury or death must be stopped, prevented or reduced.

▶ The three 'pillars' through which this is to be achieved are safer speeds, safer vehicles and safer roads and roadsides.

▶ A background requirement in all this is that road users must be 'alert and compliant'.

▶ Inputs into the system are admittance to the system, emergency medical treatment (added after 2008), education and information supporting road users, understanding crashes and risk, and legislation & enforcement of road rules. All these need to support the core objective.

▶ Mentioned under *Safer Roads & Roadsides* and *Safer Speeds* is the idea that the system needs to be forgiving of human error.

▶ The outcome is safer travel.

Let us now consider one of the major aspects of the Safe System — the elimination of fatal and serious injuries from our roads.

Chapter 2

IS ZERO POSSIBLE?

Johnston et al argue that the elimination of disabling injury and death should be a 'non-negotiable' given in all planning decisions about roads.[11]

Main Roads Western Australia's road safety strategy has the vision 'To eliminate death and serious injury crashes on the Western Australian road network', adding that 'the vision may not be achieved in the short or medium term[12]'.

The vision of the *NSW Road Safety Strategy 2012-2021* is 'working towards vision zero.' A vision based on a vision — that's insightful! 'The Safe System approach is a pathway that has an end goal of no death or serious injury occurring on the road transport network[13]'.

It is evident that every Australian jurisdiction is wedded to the belief that fatal and serious injuries (FSIs) will be eliminated, even if not just yet. The proviso is that road users will have to be 'alert and compliant[14]'. But how likely is this elimination?

One project I did in 2017 included looking at crashes in six large areas around schools, shopping centres and a park in a middle-suburban area in Melbourne. Over a five-year period, there were 144 recorded

---

11   See Johnston et al, 2017, p 37.

12   Main Roads WA, 2012, p 4.

13   Transport for NSW, 2012, p 16.

14   See Figure 1.1. Also for example see Transport for NSW, 2012, p 19.

casualty crashes. Of these, 37 were fatal or serious injury which is about double the proportion of FSI to total casualty crashes for the whole of Australia. Looking at the FSI crashes in detail, I have estimated that:

- ▶ About half can be addressed by engineering solutions, with a very good likelihood of effectiveness.

- ▶ The other half is unlikely to be solved quickly because they rely on changes to behaviour (small likelihood of it happening) or technology which is not yet in the vast majority of vehicles (principally autonomous braking if confronted by an object).

- ▶ If that technology did suddenly get implemented it would still leave over a third of the FSI crashes (13 = 35%) with no engineering solution, no speed limit solution and only a small likelihood of being solved by other means (behaviour change).

- ▶ A reduced arterial speed limit of 50 km/h could (by a generous interpretation) assist with just 8% of FSI crashes, though that drops to 6% if autonomous braking were available.

From this, I have concluded that:

- ▶ Some 50% of all fatal and serious crashes in the area will not go away until vehicle technology (autonomous braking) is widespread. Reducing urban speed limits might affect a few percent of these only.

- ▶ Once autonomous braking is in most vehicles, some 25% to 30% of all fatal and serious crashes (as recorded 2012 to 2016) are still unlikely to ever go away, even with reduced urban speed limits.

- ▶ In the areas studied (9.3 km of arterial roads, 2.9 km of distributor roads and 43.5 km of local streets), reduced speed limits would result in one fewer serious injury crash every 600 days.

So we might achieve a 75% reduction in FSI crashes. Will people say that's good enough? Well, not in NSW where the Centre for Road Safety tells us that 'any death or serious injury on our roads is one too many[15]' or in Victoria where the Transport Accident Commission (TAC) has told us 'zero is possible.' No one is saying in 2018 that it's 'good enough', following an actual 75% reduction in fatalities since the mid-1980s, so it's unlikely in the future. At some point, with road safety experts having raised expectations by saying they can completely solve the fatal and serious injury problem, someone is going to ask why it hasn't happened.

The first excuse will be that some percentage of the road users weren't alert and compliant.[16] '*Alert and Compliant*' is an Australian feature of the Safe System. We'll be back to blaming the driver.

Take some (not so) theoretical examples:

▶ The speed limit on one divided road with driveways has had its speed limit reduced from 70 km/h to 60 km/h. If someone is seriously injured after going at 120 km/h, do we say it's their fault? (Definitely). What about 100 km/h? (Probably) What about 80 km/h? (Umm) What about 70 km/h, which was (I propose for the example) a case of *driving to the conditions*? I expect the huge intellectual investment in the Safe System will result in strict lines being drawn to exclude miscreants like this person who went at 70 km/h.

▶ As our population ages, there will be increasing numbers of older drivers and pedestrians. Some are not going to be very *alert*. Do we exclude them if they have a crash? Perhaps we legislate that

---

15   *http://roadsafety.transport.nsw.gov.au/campaigns/towards-zero/index.html* (Dec. 2017).

16   As explained by Mooren et al, 2011, 'the *Safe System* approach, emerged in the State of Victoria in 2004 and was later endorsed by the Australian Transport Council in their 2004-2005 Road Safety Strategy. ... The *Safe System* approach is founded on Vision Zero principles, but requires that road users remain alert and compliant in order to ensure harm avoidance.'

all pedestrians must wear a bright, retro-reflective outer garment at night time. Then we can exclude from our thoughts those that don't comply. It was their fault.

▶ A pedestrian gets hit at mid-block pedestrian signals after crossing against the red man signal. Do we exclude her or him because they did not comply? In the pre-Safe System days, we'd seek to find the cause of the problem (i.e. do a crash investigation) and discover the long signal phasing (waiting a long time for the green walk signal) was the primary cause.

▶ At a fatal crash site (T intersection on the left) a truck driver drove straight ahead at night time and a driver exiting the side road was killed. The council had changed the intersection priority and the truck driver was supposed to give way. But the only hint was an ambiguous side-road warning sign (without advice to give way). No amount of lower speed limits will fix that problem. Do we blame the truck driver for failing to give way as the law required him to?

I predict that we will end up with a large number of casualties excluded from the picture so authorities can say that our Safe System is actually working. These casualties will disappear off to Gulag *SafeSystemThink*. 'Uncle Joe loves his people. If only they would comply and be alert, Uncle Joe could help them.'

Where have our most effective, ongoing road safety gains come from? Firstly from the early *silver bullets* of seat belts and random breath testing. Also from improved vehicle technology like safer internal design elements, then driver assistance technology. The benefits of these are well understood. What is less well appreciated and acknowledged is the major contribution of safe road infrastructure and effective road engineering solutions which are designed through an understanding of human factors, in particular road user decision-making. It has been known for a long time that, even though the most frequent contributor to crashes is road user error (i.e. human error or behavioural aspects), it is far more effective to re-engineer the road to elicit safer responses than to target the behaviour directly. Figure 2.1 (dealing with the causes of crashes) illustrates this.[17]

However, most infrastructure costs money, so the Safe System approach advises that, at least in the interim, the speed limit should be lowered if crashes are happening. Sadly, this approach is already leading to the following problems:

▶ Where the new speed limit is unreasonably low for the conditions, enforcement is required[18]. That might be considered a worthwhile community investment if the crashes are speed-related, but typically they aren't — or no one has done the analysis to see if they are or not. So enforcement effort is wasted.

▶ The future infrastructure improvements don't eventuate.

▶ Safety is viewed simply in terms of speeds and compliance, rather than as a complexity of issues.

▶ The causes of the crashes are not actually addressed. Those causes remain there, ready to create new crashes.

---

17   Also see Lydon & Turner, 2017: 'it is important to remember that though crashes often result from human error, the solutions can be found in the provision of infrastructure'.

18   Enforcement is required because most people 'drive to the conditions', meaning they drive according to their perception of what is a safe speed. If travel speeds actually do need to be lower (e.g. because drivers' perceptions of the risks doesn't match the actual risks), the physical conditions need to be altered so drivers will adopt the desired travel speed. Otherwise, for a lower limit to be effective its purpose needs to be obvious and credible, it needs to be over a short distance and such use of limits across the network needs to be rare. These days there is extensive use of low speed limits.

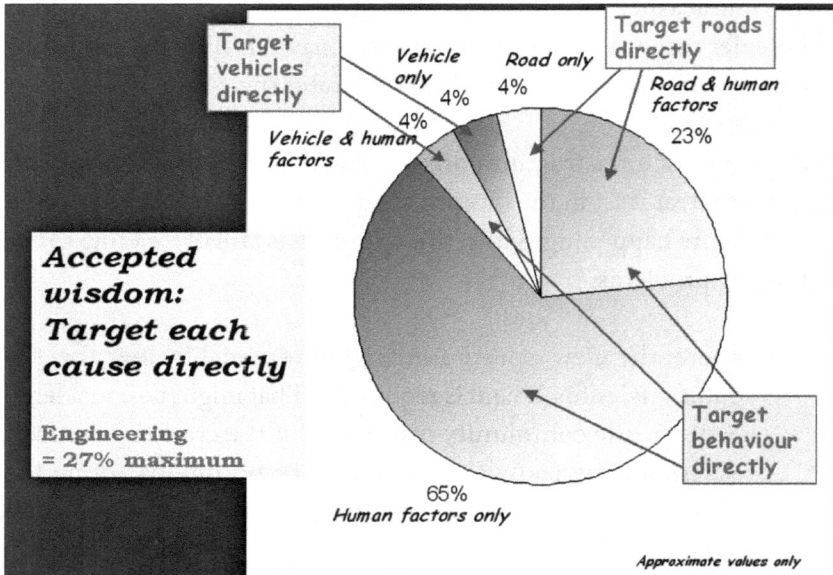

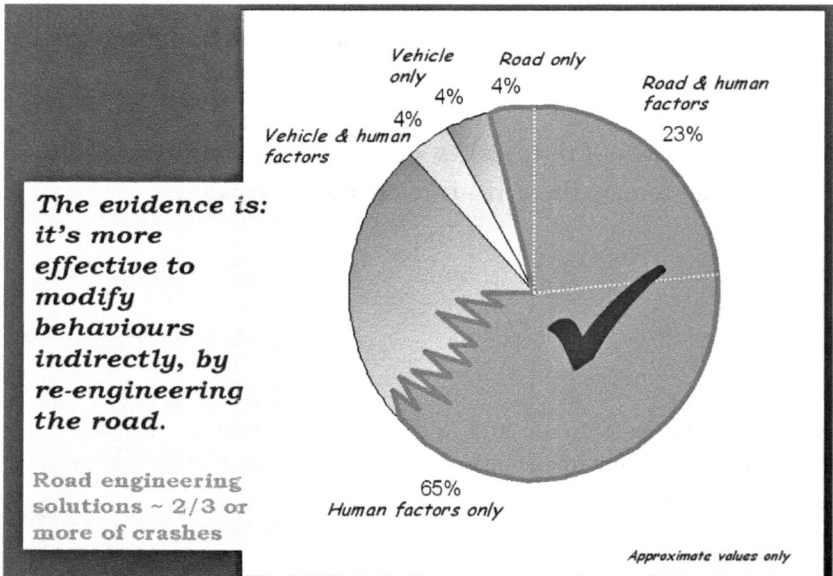

Figure 2.1
The importance of infrastructure and engineering solutions

CASE STUDY 1

No crash problems,
just a general desire to improve safety

With a major shopping street 90% bypassed some years ago, a Victorian Council sought to improve conditions by introducing a 40 km/h limit. By this time (the era of Safe System) VicRoads' speed limit guidelines had become so vague that almost any limit (high or low) could be proposed by Councils on their own roads and be accepted.[19] Council decided to implement a *40 km/h Area* limit on this road plus an area of local streets around it. Most local intersections already had roundabouts. Humps at school crossings have since been installed. My observations are:

▶ Speeds in the main street are little changed (40–50 km/h in free-flow conditions).

▶ Most people tend to drive at ~50 km/h in the local streets.

▶ Pre-existing road safety problems remain untreated: e.g. every summer Agapanthus flower heads in a splitter island completely block the view for drivers exiting the area onto the arterial bypass. Pedestrian signals in the shopping street have such a long cycle time that pedestrians cross against the red man. *Lollypop* street trees in front of the signals obstruct the traffic signal lanterns (see Figure 2.2).

▶ Some new traffic treatments in the area exhibit signs of basic cluelessness about safe traffic design. For example, at a new zebra crossing old edge lines remain, leading drivers into new islands. The 'walking legs' signs on one corner lasted a week before being knocked over by a left turning truck.

---

19   It was an engineer at another Council who raised concerns about this. Historically Council engineers have been able to quote the guidelines in order to put a stop to silly speed limit proposals from Councillors or the public. Not any more. As an aside, the draft of these guidelines was written so poorly it used the terms 'speed limit' and 'speed management' interchangeably.

Fig 2.2
Unattended-to safety issues in an area with a 40 km/h limit

CASE STUDY 2
Crashes on a 70 km/h divided road

In 2015 the speed limit on a 6.5 km length of six-lane divided road with driveways was reduced from 70 km/h to 60 km/h because it 'has a significant crash history with 199 crashes recorded in the five years ending July 2013, including 29 involving pedestrians'. When I examined the crash data in detail I established that:

▶ 60 of the 199 crashes involved no vehicle travelling along that road. Three were miscoded/unclear, two involved bicycles on the footpath and 55 only involved vehicles from side roads.

▶ The number of pedestrian crashes involving vehicles on/to/ from the road was 19, not 29. Only one of these was likely to benefit from the lower speed limit.

▶ Of the 139 crashes on the road, 20 involved going through a red light, eight were speeding or road rage, three were a side road driver (with good visibility) not giving way left turning onto the road, nine had already been treated under blackspot programs, five were a pedestrian ignoring the red man signal at intersections or crossing within 60 m of signals, 39 were where circumstances or the police details showed the speed was already 60 km/h or less.

▶ This left, at the absolute most, 55 crashes that may have benefited from the lower limit. The actual number could be half that (<1/km/year).

▶ A check of crash rates per unit of traffic volume (using 197 of the crashes) showed that only one intersection had a rate above the Victorian average: some of its crashes had been treated; most others involved only side road traffic. Four mid-block sections (out of 11) had higher than average crash rates: two had one crash type, fixable by closing median breaks, one had pedestrian signals uncoordinated with nearby intersection signals (plus anti-social driving). The fourth location is shown in Figure 2.3. Six of its seven crashes involved right turns off the arterial road colliding with a (non-bus) vehicle in the newly created bus lane while traffic was queued in the other lanes. A lower speed limit is of no use in that situation.

A road safety audit I led at the same time highlighted numerous safety hazards, including a worn-out centre line and lane lines on the only curved, undivided section of this road. Well over two years later those lines had not been re-marked. A more obvious road safety hazard would be hard to find. Nor had any other item that was advised to VicRoads been addressed, including Figure 2.3. (The road covered three local government areas; one Council did respond, cutting back trees on a side road that blocked the view to signals).

'OK, but surely the lowering of the speed limit will have a general positive road safety benefit?' perhaps you say. Without another five years' data we can't say with certainty, but consider this: this road used to have a 60 km/h speed limit (when 70 km/h was not available as an option in Victoria). With a review of speed limits in the early 1990s, 70 km/h became the new standard for divided roads with driveways. This and

Fig 2.3
Right turns through queued traffic colliding with a vehicle in the bus lane (Aerial photo courtesy of Nearmap.com.au)

other similar roads had their limit increased to 70 km/h. A study of the change by MUARC[20] drew the following conclusion:

> For the speed zone changes from 60 to 70 km/h, there were small speed increases observed on undivided roads and <u>no speed increase on divided roads</u>. Whilst <u>speed distributions narrowed on the divided roads</u>, the effects on undivided roads were less clear with some distributions widening and others narrowing. <u>No significant changes in casualty crash frequency were observed</u> for this speed zone change which is consistent with the results of speed monitoring finding little speed change and <u>mostly distributional narrowing of speeds</u>. (*my underlining*)

There is a general relationship between higher speeds and higher crash risk — Safe System material tells us this endlessly. However, a 'narrowed speed distribution' means more people drive at similar speeds, and this is also known to contribute to a lower crash risk.[21] It would appear from the Melbourne experience in the 1990s that one effect appears to cancel the other on roads of this type when the change is between 70 km/h and 60 km/h. Thus the lowering of the speed limit in Case Study 2 would appear to have been a pointless exercise.

From these two case studies I conclude that:

▶ Skills in both the state road authority and municipal Councils are so deficient these days that basic crash analysis techniques are not used or understood.

▶ The value of crash analysis (and the road safety engineering experience that builds from it) is simply not appreciated. The crashes are happening for specific reasons and they're not speed-related. What is the reason in each case?

---

20   Newstead & Mullan, 1996.

21   As long ago as the late 1960s U.S. studies established this. E.g. see Crinion, 1969.

▶ Due to 'the primacy of speed management[22]' under the Australian Safe System, speed limits are being reduced for no great effect, while the actual causes of crashes are not being fixed, despite known and effective (and generally low cost) engineering treatments being available.[23]

By my reckoning, even with effective blackspot programs and the treatment of crash problem sites, we are unlikely to ever reduce fatal and serious crashes by more than 75% over current levels. With the focus of the Safe System in Australia being 'the primacy of speed management' the results are going to be considerably worse, as the actual causes of numerous crashes will not be investigated and addressed.

Not only is zero not possible, its pursuit through the Safe System will result in:

▶ a return to blaming the driver,

▶ increasingly absurd and pointless efforts to reduce speeds,

▶ increased enforcement and penalties for no good purpose in the name of road safety (what Quayle refers to as bullying[24]),

▶ fewer safe infrastructure treatments being implemented, and

▶ the actual causes of crashes not being addressed.

---

22 OECD/ITF, 2008, Box 5.2 on p 113: 'Australia's Safe System — Illustrating the primacy of speed management'. The Editorial Group chair for this document was Eric Howard, the originator of the Safe System and one of the authors of Johnston et al, 2017.

23 This is like the fulfilment of my 2007 prediction that 'Where low speed limits are the main thrust of a road safety strategy, it is inevitable that poor design will increase and levels of safety will get worse.' (Morgan, 2007).

24 'What distinguishes bullying from other forms of intimidation is that the perpetrator is able to manipulate situations in such a way as to make it appear that the victim is only getting what they deserve.' Quayle, 2015, p 229.

Chapter 3

SWEDEN'S VISION ZERO

The origins of the Safe System lie in Sweden's *Vision Zero* and the Netherlands' *Sustainable Safety*.[25]

Vision Zero was developed in Sweden in the mid-1990s and endorsed by Sweden's parliament in 1997 as the basis of that country's road safety strategy. As illustrated by the quotes at the start of Chapter 2, this is where Australian jurisdictions' vision of zero fatal and serious injuries originated. Vision Zero has been described as a 'paradigm shift[26]' and 'the breakthrough advance in traffic safety thinking of the last decade[27]'.

So, what are the main elements of Vision Zero and just how new are they?

1. *The objective of completely eliminating fatal and serious crashes.*

This is the primary element of Vision Zero and it was indeed a new approach. In Chapter 2 I concluded that not only is zero impossible, but the methods used in the pursuit of zero will result in worse safety.

2. *Instead of focusing on reducing the number of crashes, Vision Zero's focus is on reducing and eliminating fatal and serious injuries.*

---

25 OECD/ITF, 2008, Section 5.1.2 on p 108 — though there is little evidence of the Dutch approach in the Safe System.

26 Turner & Lydon, 2017, p 598.

27 Johnston et al, 2017, p 155.

This was also a new approach, which is also at the core of Australia's Safe System. As will be explained in Chapter 5 this approach will actually be counter-productive.

3. *Under Vision Zero there is a shared responsibility for traffic safety.*

This involves both the road authorities and the road users. Road authorities are primarily responsible, as they design and manage the roads. Current wisdom is that before Vision Zero virtually all responsibility for road safety was placed on road users: it was a case of *blame the driver — or other road user* [28]. I don't know what was happening in Sweden before 1997, but in Australia state road authorities had well and truly shifted away from blaming road users by the mid-1970s. Once road safety auditing had been adopted nationally by c.1992 there was a clear acceptance by state road authorities of their responsibility to 'design for road users' safety.'

Admittedly different sectors viewed road users' mistakes differently. Victoria's Transport Accident Commission (TAC) continued for decades with advertising campaigns trying to directly change road user behaviour. But in the world of traffic management, the ideas of joint responsibility and designing the road system to accommodate the limits of human capabilities are very old. A highly regarded 1969 Australian traffic engineering manual (Clark & Pretty, eds, 1969) includes numerous items of advice and references to research on topics of how to design for road users, including the original version of the diagram in Figure 3.1, which was first published in 1964. It illustrates how the limits of a driver's ability to process information can lead to poor decision making.

---

28   E.g. see Belin et al, 1997.

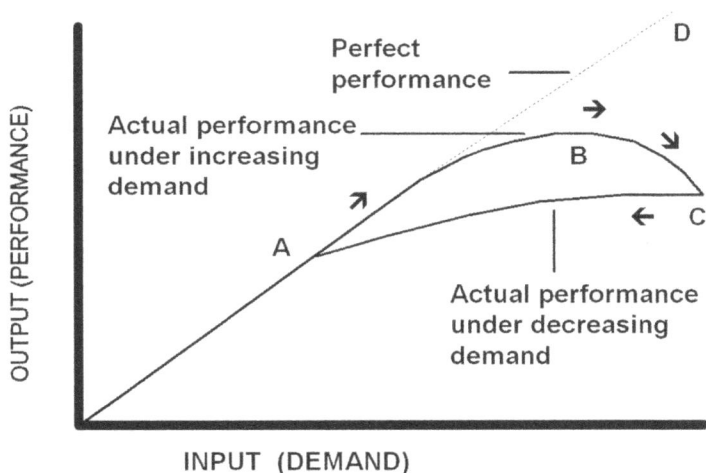

Figure 3.1
Information Processing Model[29]

By the time Victoria's then-Transport Minister Jim Kennan stated in October 1989 that 'Most road accidents are caused by bad driving', he was viewed with derision by most professionals.

In 1992 the Victorian branch of the Institute of Municipal Engineering Australia, with VicRoads assistance and funding, produced the Victorian Local Government Road Safety Strategy. Its front cover is shown in Figure 3.2 and clearly shows the idea that road safety is a joint responsibility.

---

29   Based on the model in Cumming, 1964 and also illustrated in Cumming & Cameron, 1969. (Courtesy of Australian Road Research Board).

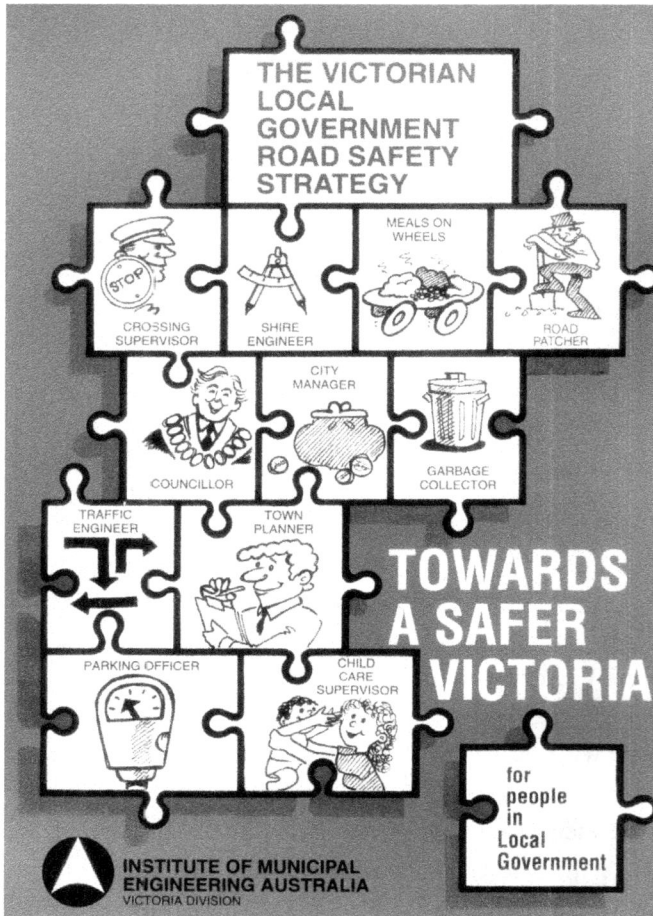

Figure 3.2
Front cover of the 1992 Local Government Road Safety Strategy
(Courtesy of Institute of Public Works Engineering Australasia, Victoria Division)

I spent the 1980s in local government designing for road users and investigating blackspots, then designing and implementing remedial treatments that helped road users make safer responses. Should I have been blaming road users instead?

Belin et al, 1997 suggest that because Sweden's Road Traffic Ordinance required road users to 'exercise ... attention and caution' and 'show respect for other road-users' this equated to the entire responsibility for road traffic safety being placed on the individual road user.[30] Australian Road Rules (and predecessor regulations) contain similar obligations for road users to keep a look out and act reasonably, etc.. If Belin et al's assertion was true road authorities would never have been found responsible in courts of law for injuries sustained in crashes. Yet over the past 50 years road authorities in Australia have increasingly been found negligent in the ways they have designed or managed roads. In numerous cases, the *blame* has shifted away from the individual road user.[31]

Clearly Vision Zero has not been responsible for introducing a *joint responsibility* approach to road safety and a shift away from blaming the road user. Those shifts had been going on for decades.

4. *Humans make mistakes. Vision Zero advises that the road system should be designed to be forgiving of those mistakes.*

The acceptance that road users make mistakes is not new. For example, the concept of Clear Zones is based on it. The new concept Vision Zero has introduced is that there should be a limit on the consequences of those mistakes: no fatal or serious injury. In other words, the new element is that this objective is clearly spelt out. So, whereas a Clear Zone standard is based on approximately 85% of errant vehicles (and occupants) being saved from death or injury, Vision Zero says 100% of them should be 'saved'. The idea that (road design) standards do not equal safety is of course not new. For example, the first edition of Australia's Road Safety Audit guidelines in 1994 states 'strict adherence to standards does not always result in the safest possible design[32]'.

The benefit of the Vision Zero approach (design so that fatal and serious injuries are avoided) is that it clearly states the design parameters. Once it is adopted in jurisdictions' strategies, it theoretically makes it easier

---

30 Belin et al, 1997 at Section 2.1.2.

31 It is possible that the development of case law (e.g. with misfeasance and nonfeasance) and subsequent legislation has played a greater part in Australia than in Sweden.

32 Austroads, 1994 at Section 10.

for designers to implement more expensive — and more effective — design solutions. Of course, the risk is that the vision of zero deaths and fatal injuries becomes the new standard which is strictly adhered to (see Chapter 2 for the consequences of that).

5. *Where the existing infrastructure is unable to provide the required safeguards against fatal and serious injury, the speed limit must be lowered.*

In other words, there can be no balancing of mobility and safety: safety is not a variable in the equation — it is paramount. As described by Professor Claes Tingvall[33] this initial reduction of speed limits (say from 100 km/h to 80 km/h on undivided rural roads) will lead to road investment decisions being made on the basis of whether the investment required for the necessary improvement in safety (so 100 km/h can be restored) is worthwhile in order to achieve the desired level of mobility.

With our vast, dispersed network of rural roads, the question for Australia is this: if the default rural speed limit is lowered in line with Safe System principles (say to 80 km/h, as in Sweden) would a satisfactory benefit/cost ratio (BCR) be achieved for safety improvement projects to allow the majority of the network to return to existing speed limits — or even to 90 km/h? The answer is definitely 'No': Australia has over 700,000 km of rural roads, with two-thirds of them unsealed. If wire rope barriers along the centre and both sides of a road cost $1 m/km, this cost — plus formation widening for installing barriers to prevent head-on and run-off-road crashes — would be in the order of $1,500 billion across the whole rural network, which is seventy-five times the nation's total annual road infrastructure investment. Even if just the top 25% of the rural network was treated and a whopping 20% of all road infrastructure spending went on Safe System barriers (ignoring the BCRs) it would take almost 100 years to bring into effect. So most of the roads would stay at 80 km/h and either:

---

33   Then the Director of Traffic Safety at the Swedish National Road
      Administration, as quoted by Johnston et al, 2017 on pp 80-81.

- ▶ there would be a heavy investment in speed cameras (which still could not be everywhere), or

- ▶ speed variation would increase as some drivers stuck to 80 km/h and others 'drove to the conditions', reducing any safety benefits of the lower limit. This is already common on 80 km/h rural and semi-rural roads.

Johnston et al comment that 'enforcement [is] least effective on such [rural] roads because of the vast lengths of road to be covered[34]'. So this is certainly an original idea with Vision Zero, but the implications for Australia are, frankly, absurd.

Also, see Chapter 6 for discussion on making mobility subservient to road safety and the implications of this for other values societies have.

6. Technologies.

The Trafikverket.se (Swedish Transport Administration) website (accessed late December 2017) describes under *Key innovations based on Vision Zero*:

> Median barriers and cameras are innovations that have increased the level of safety on Swedish roads. Older solutions such as roundabouts and alcolocks [ignition interlock devices] have been developed and have acquired greater importance since Vision Zero was established.

Let me put to rest the adopted wisdom that any of these treatments occurred because of *Vision Zero thinking*. The '2+1' schemes with a median barrier were being developed in Sweden from the early 1990s, in response to high crash rates on 12-13 m wide undivided rural roads built mainly between 1955 and 1980. These roads were marked with a centre line, but were effectively three lanes wide, with the central

---

34 Johnston et al, 2017, p 65.

area used for overtaking[35] (see Figure 3.3). The first '2+1 with barrier' scheme opened in 1998.[36]

Meanwhile, the Vision Zero concept was being developed at the same time and was first presented publicly in 1995. It was adopted by the Swedish parliament in 1997. It is evident that the development of '2+1 with barrier' schemes is not the result of *Vision Zero thinking*: it developed either before or coincidentally, as an engineering response to an identified serious crash problem that resulted from an earlier design concept (13 m wide rural roads) that could not operate safely with the higher traffic volumes that arose.

It should be noted that '2+1 with barrier' treatments are sub-optimal: there are almost 2,000 median repairs per year[37] in Sweden and fatal and serious injury (FSI) crashes have been reduced by 39 — 63%, rather than being eliminated or substantially eliminated.[38] But they are much cheaper than full-standard duplication and for Sweden, they avoid archaeological delays.[39] Elsewhere in the world (where zero is not the target) they would rightly be regarded as a road safety success, but why so in Sweden?

---

35 Australia thankfully had few of these, mainly some urban roads in the 1950s and 1960s. Marked either as two lanes or three lanes, the central part was used for overtaking, leading to head-on and other related crashes. At an early date these roads were re-marked to provide standard width lanes dedicated to traffic in a single direction. A survey in the UK in the early 1960s showed that casualty crashes dropped on average by over 20% and fatalities halved where two-lane roads were widened to three lanes, though on 20% of widenings crashes increased (Leeming 1969, pp 46-47).

36 Bergh et al, 2016.

37 Ekman, 2014, slide 25.

38 Carlsson, 2009.

39 'Whenever a road or intersection is to be widened in Sweden, an archaeological survey is required. There is so much historic Viking material in the ground throughout Sweden that any road widening is likely to be held up for years. So the incentive is high for keeping within the existing road width.' (Morgan, 2007)

Figure 3.3
Risky overtaking on a Swedish 3 lane wide road

Consider also the following elements in what I've called the *Safe System Manifesto* (see Chapter 1):

▶ Life is sacrosanct, and

▶ There can be no compromise: zero is not negotiable

Compromises have been made with the introduction of '2+1 with barrier' treatments in Sweden; life has not been sacrosanct — and some great gains have consequently been achieved. Evidently, the Swedes are taking a more pragmatic approach to achieving zero, compared with the proponents of Australia's Safe System.

As for the other three 'innovations based on Vision Zero', surely no one is suggesting that speed cameras (1980s), roundabouts (1905) or alcolocks (ignition interlock devices — 1980s) were waiting for Vision Zero to come along before practitioners started using them?

All of this raises a theme that I will develop in more detail later: that in fact much of what is successful in Sweden is not new; it's good old road safety engineering — providing safer infrastructure, based on an understanding of crash causes. It's not a 'new paradigm' or a 'new way of thinking', at least it wouldn't be in Australia. Maybe it's new

Figure 3.4
Part of the reason why '2+1 with barrier' roads don't have zero fatal and serious injury crashes? Worrying intersection layouts

in Sweden where by around 1990 they were placed at a disadvantage through having extensive lengths of hazardous road layouts (i.e. their standard 13 m wide single carriageway roads) that, in my view, resulted from an inadequate history and culture of good road design.[40]

In summary, the new elements that arose with Sweden's Vision Zero are:

▶ The objective of completely eliminating fatal and serious crashes. This is not actually achievable, as discussed in Chapter 2.

▶ Instead of focusing on reducing the number of crashes, Vision Zero's focus is on reducing and eliminating fatal and serious injuries. This approach is counter-productive, as discussed in Chapter 5.

▶ The spelling out of a clear objective (zero fatalities and serious injuries).

▶ Mobility and safety must not be balanced or traded off. Safety

---

40   See Morgan, 2007 and the end of Chapter 5 for a discussion of the reasons for this view.

(sufficient to achieve zero fatalities and serious injuries) must be provided. The implications of this for Australia are, in my view, absurd. Prior to Vision Zero others, including Haight in 1994, had already exposed this one-dimensional approach to road safety as being at odds with what societies actually believe — see Chapter 6.

The other elements now viewed as part of Vision Zero were pre-existing. That's not to say these other elements are not worthwhile or even essential for improving road safety. Indeed, if it takes some charming, blonde-haired people from some mystical other place to re-badge previously known engineering solutions and say that 'We are going much more for engineering than enforcement[41]' and if that stops all the recently self-appointed Safe System experts in Australia who think that road safety success is about directly changing or controlling human behaviour (without compromise), then more power to those Swedes. But perhaps that's not all the Swedes are saying.

In a media interview in New York in 2014 (Goodyear, 2014) Dr Matts-Åke Belin, from Trafikverket, Swedish Transport Administration is quoted as saying:

> I will say that enforcement plays, of course, a role in Sweden, but not so much. We are going much more for engineering than enforcement.
>
> We are doing [camera enforcement], but in a different way. ... we have put them on most rural roads. We have one of the largest safety camera systems in the world, per population. But they are not catching people — it's nudging people. ... and in a friendly but firm way, we say you have to keep [to] the speed in this area because we have a history of crashes.
>
> And we have increased the compliance on these roads from 50 to more than 80 or 90 percent. And we don't catch any people at all.

---

41   see Goodyear, 2014.

We reduce the speed, but we don't catch people. And we don't earn any money.

If engineering and infrastructure upgrades are going to be sub-optimal (as Sweden's '2+1 with barrier' treatments are) or if they are not effectively targeting known crash causes, then there is a limit to what they can achieve. Reductions in deaths and serious injuries are going to taper off. And this is what's happening in Sweden (see Chapter 4). A recent report (ITF, 2017) provides advice received from Sweden that 'speeding remains a major problem[42]' and:

Compliance with speed limits remains at an unacceptably low level. In 2014, the share of traffic volume within speed limits was estimated at 47% (target: 80%) on national roads and 63% (target: 80%) on urban roads.[43] Data for 2016 show that compliance has decreased for national roads (44%), but increased slightly on urban roads (67%).

At the end of 2016, there were around 1 500 speed cameras on the rural network in Sweden. For the period 2017-20, yearly additions of about 200 new cameras are planned. Additional use of speed cameras is especially important on roads with speed limits of 80 km/h since these roads usually do not have median barriers and speed compliance is low.[44]

It is also especially important I suggest because these 80 km/h roads probably look like 90 km/h or 100 km/h is an appropriate speed for the conditions. Sweden's road to not-exactly-zero will have some 2,300

---

42 This is consistent with my observations in Sweden in 2006 when I felt I was the only one sticking to rural speed limits. Locals would overtake me, then slow down as they passed each speed camera. There was clearly no ownership of Vision Zero by the majority of Swedish motorists. This was at a time when many speed limits were already 10 km/h lower than in Norway and Denmark, but before a further 10 km/h reduction was introduced (now implemented).

43 This is far lower compliance than the 'more than 80 or 90 percent' Dr Matts-Åke Belin claimed.

44 OECD, 2017, pp 512-513 and 517.

rural speed cameras by 2020. There's a puritanical need for control in all this[45] and somehow this is resonating with the proponents of Australia's Safe System (e.g. see Johnston et al, pp 138-139).

By contrast, the Netherlands' Sustainable Safety program — introduced in the early 1990s — is based on five principles that look far more pragmatic and acceptable (see Table 3.1): treating road users with respect and using engineering techniques to achieve safer road user responses. Rather than basing speed management on low speed limits and automated camera enforcement, the Dutch developed the concept of the 'self-explaining road'. Put simply, match the layout to the speed you want: if you want people to drive at 80 km/h (or 30 km/h), design the road and construct features within it that result in drivers adopting that speed. Then enforcement is about dealing with the aberrant few, not control of the majority.

---

45 Although looking at the figures for exceeding the speed limit in Sweden there doesn't appear to be an acceptance of being controlled, like there is in Victoria.

Table 3.1
The five Sustainable Safety principles (Netherlands)[46]

Sustainable Safety principle	Description
Functionality of roads	Mono-functionality of roads as either through roads, distributor roads, or access roads in a hierarchically structured road network.
Homogeneity of mass and/or speed and direction	Equality of speed, direction, and mass at moderate and high speeds.
Predictability of road course and road user behaviour by a recognizable road design	Road environment and road user behaviour that support road user expectations through consistency and continuity of road design.
Forgivingness of the environment and of road users	Injury limitation through a forgiving road environment and anticipation of road user behaviour.
State [of] awareness by the road user	Ability to assess one's capability to handle the driving task.

This makes so much more sense than the high-minded rhetoric of Vision Zero and the Safe System which both inevitably lead to low speed limits and automated speed camera control, while actual causes of crashes are not properly addressed, as illustrated in Chapter 2.

As in Australia with the Safe System, the future of Vision Zero is going to see a battle between the infrastructure/re-engineering approach and the approach of low speed limits with stronger, automated enforcement. With the flattening out of crash-related rates (such as fatality rates shown in Figure 4.3) and the never-ending need to look

---

46 SWOV, 2013. Note that the Safe System does not embrace 90% of these
    principles.

for new approaches to achieve the unachievable zero, it's inevitable that Vision Zero will turn into a system of ever-more rigorous automated speed control. How else can 80 km/h travel speeds by the vast majority of drivers be achieved on roads where 90 or 100 km/h looks like an appropriate speed for 'driving to the conditions'?

The *Trafikverket* Swedish Transport Administration website includes a page that explains Vision Zero.[47] At the top of the page is a photo of a car travelling past a Swedish speed camera. Perhaps the connection was unintended but directly under it is written 'This is Vision Zero'.

My request to reproduce the *This is Vision Zero* web page got lost somewhere in the bowels of the Swedish Transport Administration. So instead, Figure 3.5 is an image I took in 2006.

Figure 3.5
Swedish speed cameras at a rural intersection
where the speed limit has been reduced

Figure 3.5 shows speed cameras at a rural intersection in Sweden where the speed limit through the intersection had been reduced from 90 km/h to 70 km/h, evidently as a crash reduction measure. It was apparent from inspection that at some times of the week, long queues built up on one side road approach and two lanes formed, though only

47   *www.trafikverket.se/en/startpage/operations/Operations-road/vision-zero-academy/This-is-Vision-Zero/* (Jan. 2018)

one lane was provided. Was this an indication of inadequate gaps in main road traffic, leading to crashes? Whatever the reason behind the 70 km/h speed limit, it was evidently not effective because by 2011 Google Street View shows the speed limit has been further reduced to 60 km/h (while the approach 90 km/h limit is unaltered). This does not solve the crash problem — it simply reduces the consequences when incidents or accidents happen. To solve the problem, money would need to be spent on an effective remedial engineering treatment.

As the Swedish website says, 'This is Vision Zero'. Smile!

Chapter 4

RACE TO THE BOTTOM

Graphs are available showing the relative positions of nations with their rates of crashes (e.g. see Figure 4.1).

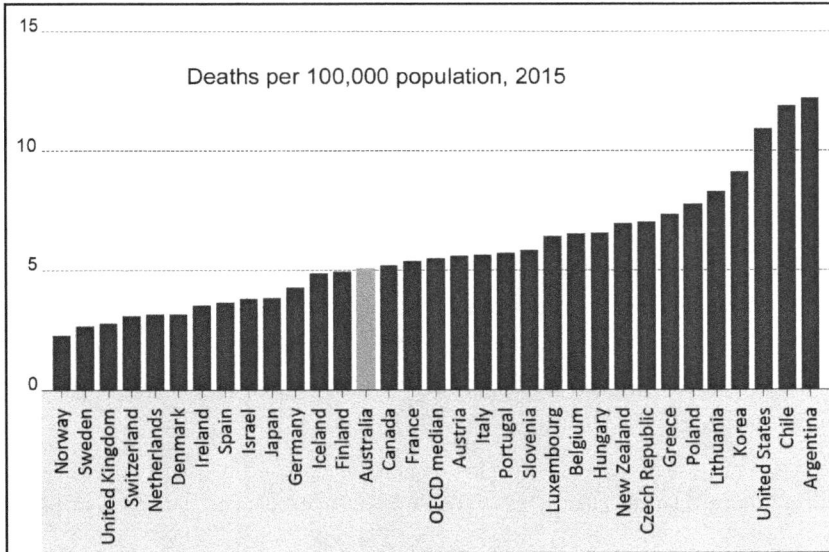

Fig 4.1
International comparison of fatality rates
per 100,000 population[48]

Inevitably these are used to berate a jurisdiction's efforts and make the case that they need to do better. For example, Truong and Cockfield, 2015 stated:

---

48   BITRE, 2017, extract of Figure 1.1, p 3 with text added. (Courtesy of BITRE)

Despite achieving record low road death results year on year between 2005 and 2013, Victoria has fallen well behind on the world platform. In 2003, Victoria ranked 5th with 6.7 deaths per 100,000 population, when compared to 24 other OECD nations. However, by 2012, Victoria has slipped to 16th place, with 5.01 deaths per 100,000 population.

The current Victorian Road Safety Strategy aims to achieve a 30% reduction in fatalities and serious injuries between 2013 and 2022. If Victoria was to reach its target reduction for fatalities by 2022, this would equate to 3 deaths per 100,000 and still place Victoria more than a decade behind some of the best-performing jurisdictions in the world. These countries include:

United Kingdom — 2.8 in 2012
Norway — 2.9 in 2012
Sweden — 3.0 in 2012
Denmark — 3.0 in 2012

It is clear additional actions needs [sic] to be taken to help Victoria further reduce road trauma. The fact that other countries can achieve such positive results encourages the TAC to better understand and learn from the experience of some of the world's best performers in road safety.

The Netherlands is another favourite to quote (3.4 in 2012 — *dat kan beter!*[49])

Yet in 2012 the fatality rate in the Australian Capital Territory was 3.2 — and 1.8 the next year (per 100,000 population). Did anyone say 'Cancel my study trip to Europe, I'll go to Canberra instead'? Johnston et al comment (p 30) on 'invalid comparisons across disparate problems' and the need to disaggregate data. I agree. When Australian data by jurisdiction and by degree of remoteness are perused it is fairly obvious that the major factor in lower crash rates is the degree of urbanisation.

---

49   Just kidding. Heel erg goed !

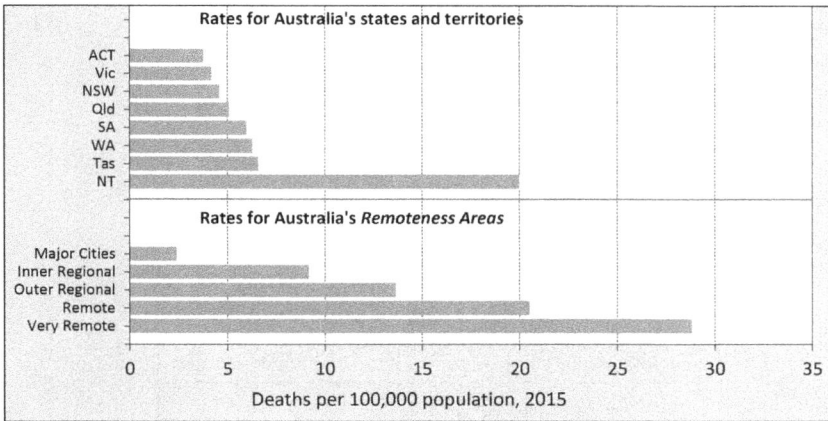

Fig 4.2
Fatalities per 100,000 population, Australia[50]

This is consistent with examples in Europe: the UK has 65 million people in an area similar to Victoria (which has one-tenth the population). The Netherlands is the most densely populated country in the world. Another factor is *motorisation* (car ownership[51]). Rates in Europe are much lower than in Australia: see Tables 4.1 and 4.2.

Most population growth in Australia is in its two biggest cities where options for public transport are increasingly available. Elsewhere we have to work with the car ownership rates we have.

After that, crash rates will relate to the level of physical safety of the road infrastructure, all other things being equal. In the Netherlands 'about 40% of travel occurs on motorway standard highways[52]'. Since 1998 Sweden has been converting three-lane-wide rural highways to '2+1' highways with a physical barrier separating opposing traffic. By

---

50   BITRE, 2017, extract of Figure 1.1, p 3. (Courtesy of BITRE)

51   Although car ownership levels will also relate to levels of urbanisation: the more urbanised, the more opportunities for travel by other modes and the less need for your own car.

52   Johnston et al, 2017, p 30.

2025 they plan to have median barriers on all roads 80 km/h or higher which have 2,000 veh/day.[53]

Table 4.1

Motor vehicles per 1,000 population[54]

Country	No. of motor vehicles per 1,000 population
U.S.A. (2015)	795
New Zealand (2017)	774
Australia (2017)	740
Canada (2014)	662
Japan (2017)	591
Norway (2014)	584
Netherlands (2010)	528
Sweden (2010)	520
U.K. (2010)	519
Denmark (2010)	480

There are a few points I make from all this. Firstly, in many parts of Australia we may not be doing as poorly in road safety as proponents of zero fatal and serious injury crashes are suggesting — so far as fatality rates are concerned. It is possible that Sweden will always have a lower national fatality rate because it has lower car ownership and because its dramatic early reduction in fatality rate (2000 to 2010) was largely due to replacing inherently unsafe infrastructure (three-lane roads) with much safe infrastructure (2+1 with a median barrier).

---

53   ITF, 2017, p 517.

54   *https://en.wikipedia.org/wiki/List_of_countries_by_vehicles_per_capita* (Feb. 2018). Yes, I appreciate it's only Wikipedia, but you get the general idea.

Table 4.2

Cars per 1,000 population, Australia[55]

State / Territory	2009	2013	2014
Australia	730	749	756
Tasmania	798	853	861
NSW	652	678	683
Victoria	755	771	774
Queensland	768	782	790
South Australia	757	781	791
Western Australia	828	826	840
Northern Territory	579	623	627
ACT	703	725	727

Secondly, looking at the rates year on year it is evident that getting ongoing reductions in crash rates is actually difficult, if not impossible. Invariably new initiatives show good results for a limited time, then the effects plateau out. There is no automatic trend line to zero, just because we wish it to happen. So, for example, tripping over ourselves to look Swedish is fraught with long-term hazards for the credibility of road safety efforts in Australia.

Sweden's fatality numbers (and rates) have plateaued since 2010 (see Figure 4.3).

A report to the Swedish parliament in 2017 advised that 'it will be difficult to reach the 2020 target both for fatalities and seriously injured if no measures beyond those already planned are being implemented[56]'.

---

55   *http://www.abc.net.au/news/2014-07-30/tasmanians-own-more-cars-per-capita-than-any-other-state/5636102* reporting Australian Bureau of Statistics data (Feb. 2018).

56   ITF, 2017, p 516. The targets for 2020 were halving the fatalities experienced in 2007 and reducing serious injuries by a quarter.

The Netherlands has experienced similar problems: fatality numbers and rates have plateaued.[57]

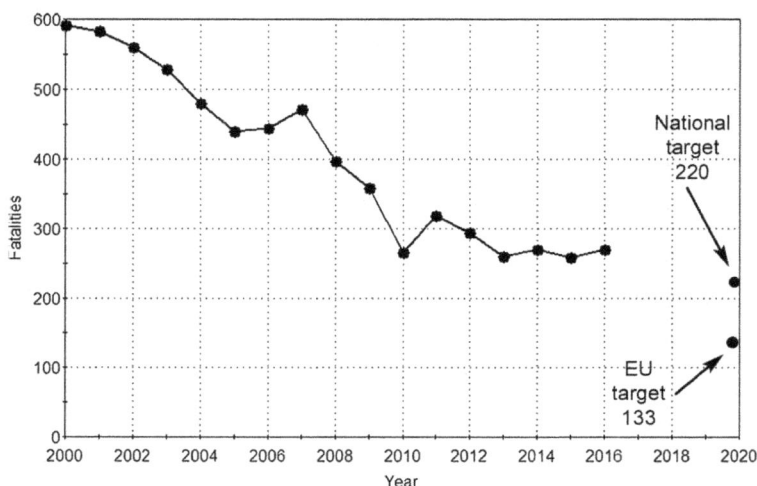

Figure 4.3

Trends in road fatalities towards the national target (Sweden)[58]

Thirdly, it's not just about fatalities. In the Netherlands since 2006, the number of serious injuries has increased by over 30%. In the UK Proctor, 2017 reports that in the period 1999 to 2015, although serious injury crashes ('Stats 19') reduced by 44%, there was a 4% increase in those serious injury crashes that are considered 'clinically seriously injured' (MAIS 3+).[59] These are the types of crashes the Safe System seeks to eliminate.

It is simply not easy.

Fourthly, we need to spend less time looking at aggregated rates and more time examining the details. Where are the rates higher? Where are the numbers higher? Which road users are most involved? What

---

57   ITF, 2017, Figure 27.5 on p 381.

58   Fatality numbers from ITF, 2017; https://www.statista.com/ statistics/438009/number-of-road-deaths-in-sweden/; https://www.trafa.se/ en/road-traffic/road-traffic-injuries/.

59   Excludes AIS codes 1 (e.g. superficial lacerations) & 2 (e.g. fractured sternum/breastbone).

road types are most involved? Etc.. Success requires putting effective countermeasures in place to target specific, identified problems. This is why the blanket application of 40 km/h speed limits around all schools was such a stupid idea: there was no blanket problem. In fact, there was virtually no problem at all around schools, anywhere. So money was wasted on all the static and electronic signs and police time is wasted on enforcing a restriction that, in most locations, has no safety benefit. The details do not support the action.

The final point I take from the examination of crash rates is that infrastructure designed specifically with safety in mind plays a major role in bringing down crash numbers (and therefore rates). It's not just the standard Safe System line about roads and roadsides that are forgiving of human error. It includes designing (or re-engineering) roads to remove conflicts and remove the need for road user decisions that lead to conflicts. The importance of safe infrastructure in reducing crash numbers and severities is generally undervalued in Australia. Consider all the right-angle crashes at crossroads that have been eliminated by installing roundabouts. In so many other ways the driving task is now much easier than fifty years ago: no give way to the right rule (uncontrolled intersections), thousands of traffic signals for safety and access, separate right turn lanes at major intersections, linked signals, divided roads, interconnected freeways, better access control, more marked traffic lanes, etc.. So perhaps part of the reason Australia has plateaued in its efforts relates to the amount of money being applied to safer infrastructure. Case Study 2 in Chapter 2 illustrates how a loss of skills in looking at crashes led to a failure to identify simple infrastructure improvements that would eliminate many crashes.

Maybe the only value in comparing Australian fatality rates with overseas rates is if it results in more money being provided to build safer infrastructure into our roads. If it results instead in money being spent on 2,300 rural speed cameras in each state it would be in line with Johnston et al's desire for intense speed enforcement[60], but to me, it would not look like the road to utopia.

---

60   E.g. see Johnston et al, 2017, pp 139 and 141.

Chapter 5

FOCUSING ON FATAL AND SERIOUS INJURY CRASHES

The focus of the Safe System is the elimination of fatal and serious injuries. By way of example, Main Roads Western Australia's strategy *The Road Towards Zero* illustrates this with the pyramid diagram in Figure 5.1. It's obvious where they wish to focus.

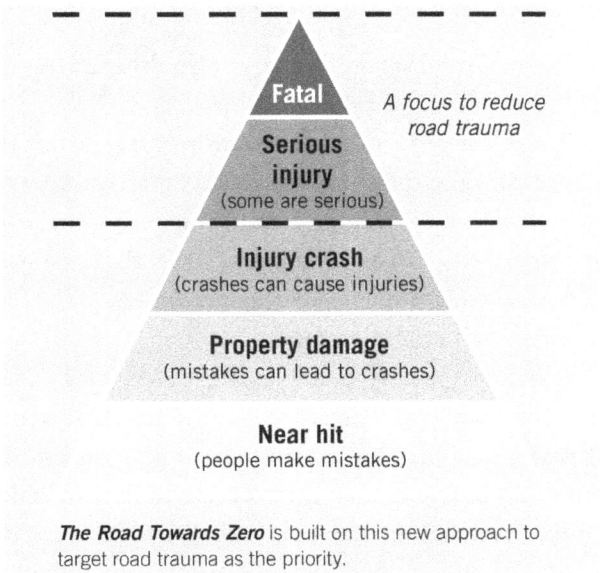

Fatal

A focus to reduce road trauma

Serious injury
(some are serious)

Injury crash
(crashes can cause injuries)

Property damage
(mistakes can lead to crashes)

Near hit
(people make mistakes)

The Road Towards Zero is built on this new approach to target road trauma as the priority.

Figure 5.1

Focus on Eliminating Fatal and Serious Injury Crashes in the Main Roads W.A. Road Safety Strategy[61] (Image courtesy of Main Roads W.A.)

---

61   Main Roads WA, 2012, p 4.

However, fatal and serious injury crashes are only a small part of the total road safety/vehicle collision problem. If only those serious injury crashes that result in ongoing disability are considered, it is an even smaller part of the problem, as Figure 5.2 illustrates.

By looking at actual numbers in Figure 5.2 it can be seen that the Safe System is concerned, not with the rather substantial-looking top segment of the W.A. pyramid (Fatal and Serious Injury), but with less than one percent of all crashes on our road network. If you are involved in one of the 600,000-plus crashes per year in Australia that make up the remaining 99% which don't result in death or an ongoing disability, the Safe System is not interested in you. You're part of the Safe System success story. Think about it — the whole focus of the Safe System is that little shaded 'Area of concern' near the top of the sixth column of Figure 5.2 and getting people out of it and down into the areas of less severe crashes or no crash happening at all.

This is our national road safety strategy — looking at 1% of crashes.

Of course, if it is viewed in terms of crash costs, rather than crash numbers, these fatal and disabling crashes account for a very large 65% of the costs to the community from road crashes — including loss of earnings, medical costs, lost production, etc.. Yet that still leaves 35% of the costs of road crashes outside the area of concern of the Safe System and our national road safety strategy.

The second edition of the Austroads' Guide to Treatment of Crash Locations provided a practical example about how the extent of a rear-end left turn crash problem at an intersection in Perth would not have been identified if *property damage only* crashes had not been included in their database.[63] In fact, in Victoria, we have had to put up with this data deficiency since 1990. Consequently, practitioners think that if it's not in the crash data, the problem doesn't exist. And now with the Safe System, the idea will develop that if it's not a serious crash it doesn't matter.

---

63   Austroads, 2004, p 28.

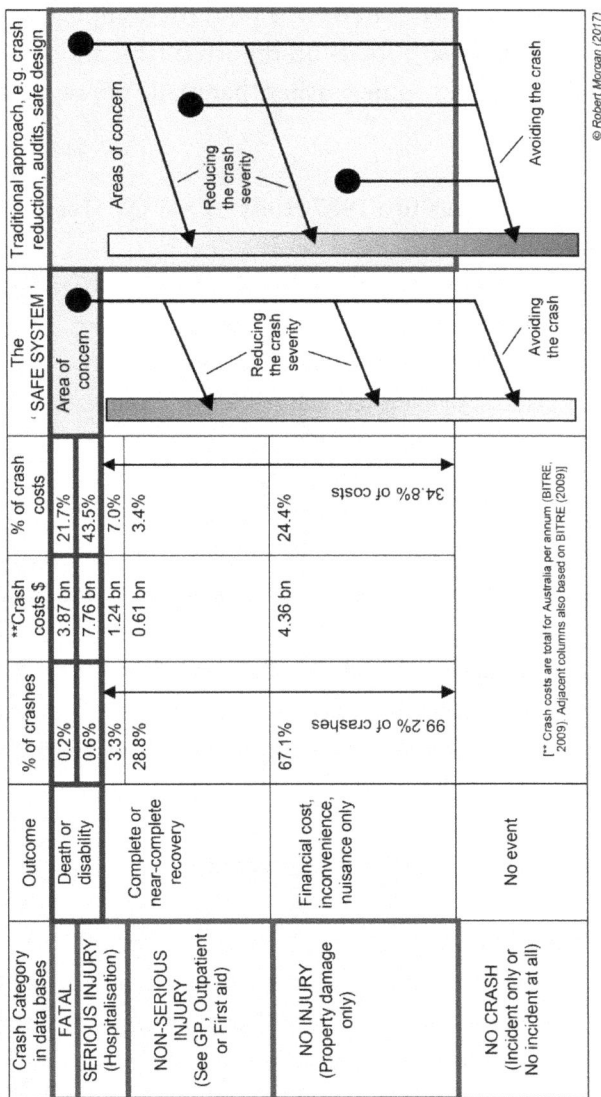

Crash Category in data bases	Outcome	% of crashes	**Crash costs $	% of crash costs	The 'SAFE SYSTEM' — Area of concern	Traditional approach, e.g. crash reduction, audits, safe design — Areas of concern
FATAL	Death or disability	0.2%	3.87 bn	21.7%		
SERIOUS INJURY (Hospitalisation)	Death or disability	0.6%	7.76 bn	43.5%		
NON-SERIOUS INJURY (See GP, Outpatient or First aid)	Complete or near-complete recovery	3.3%	1.24 bn	7.0%		
NO INJURY (Property damage only)	Financial cost, inconvenience, nuisance only	28.8%	0.61 bn	3.4%		
NO CRASH (Incident only or No incident at all)	No event	67.1%	4.36 bn	24.4%		

99.2% of crashes — 34.8% of costs

[** Crash costs are total for Australia per annum (BITRE, 2009). Adjacent columns also based on BITRE (2009)]

© Robert Morgan (2017)

Figure 5.2
Areas of concern: the Safe System vs Traditional Approaches[62]

---

62   While Safe System material typically refers to 'fatal and <u>serious</u> injuries', injuries recorded as 'serious' simply mean 'admitted to hospital'. A majority of those hospitalised fully recover in a relatively short time and are not the concern of the Safe System. The Safe System is concerned with eliminating deaths and ongoing disabling injury from our road network.

Turner, Lydon et al, 2017[64] make the point that 'injury levels are not accurately reported', so a focus on reported fatal and serious injury crashes is likely to miss some crashes that really are serious and need attention.

Andreassen, 2013 refers to a 1987 study by SWOV (Netherlands):

> While the distribution of injuries is head/neck 45%, thorax/abdomen 20% and limbs/back 35%, the distribution of disabilities is limbs/back 62%, head/neck 30% and thorax/abdomen 8%. …. The study found an overall disability incidence was 39%. For inpatients, it was 55% and outpatients was 29%. The disability problem seemed to be, to a large extent, a problem amongst slightly injured, 83%.

Andreassen then goes on to conclude 'These results suggest that all levels of treated injury need to be considered, not just serious injury'.

It has long been appreciated that the difference between a minor injury and a catastrophic injury can be a split second: for example, does a side impact occur with the front bumper bar of the car or is it into the driver's compartment?

The case for considering all crashes is compelling.

Johnston et al, in support of the Safe System, allocate a whole chapter to the idea that road safety should be approached in a way similar to preventive medicine. This includes supporting a population-level strategy whenever a low-level risk is spread widely across a community. For example, referencing work by Dr Geoffrey Rose in 1981, they say that a 5% average decrease in blood pressure across the population will have a far greater effect on reducing the number of strokes, than a program aimed at identifying those people with medically-defined

---

64   Turner, Lydon et al, 2017, p 577.

hypertension.[65] The universal applicability of this proposition in medicine and its applicability to other areas of life has been questioned by others (see the end of this Chapter). Nonetheless, Johnston et al endorse it and it therefore follows (though it is not suggested by them) that a road safety strategy that doesn't simply focus on fatal and serious injury crashes as the Safe System does, but seeks to address all crashes is more likely to succeed in reducing the total number of fatal and serious crashes. And given the uncertainties in the recording of crash severity levels, this has to be true.

The level of safety of our road infrastructure depends heavily on the level of expertise amongst practising road safety engineers. Road safety engineers gain their skills and knowledge through experience looking at accident blackspots and understanding the connection between how road elements are laid out and/or operated and the types of crashes that occur. This understanding is then used in road safety audits and other tasks that support the safe design of roads. Newcomers to road safety engineering are not going to gain much worthwhile experience and understanding if all they look at is one percent of the crashes that are happening. The implications for future road designs in Australia would be extremely worrying. How, for example, might a future road safety engineer, robbed of experience from looking at all crashes, respond to a high-speed ramp design like that in Figure 5.3?

---

65   Johnston et al, 2017, p 105. The argument is that, this being the case in medicine, then in road safety a small drop in travel speeds by everyone will be more effective in reducing crashes than only targeting obvious 'speeders' and getting them to slow down.

Figure 5.3
A typical Swedish entry ramp onto a high-speed road

Figure 5.3 shows a typical Swedish entry ramp design where the entry angle increases the risk of an entering vehicle crossing into the median-side lane or hitting the median barrier at a high angle. In fact since that photo was taken in 2006, box beam median barriers have been installed opposite similar entry ramps in Sweden to counter the problem.[66] Contrast that design with the long-taper, narrow-angle entry ramps typical in Australia, the UK and Denmark. In 2007 I commented that designs like this indicate the absence of 'a history and culture of good road design' (Morgan, 2007). It was my conclusion by then that it has been a history and culture of good road design within road authorities in places like the UK and Victoria that enabled the swift introduction of road safety auditing. This was because good road design included monitoring crashes and identifying where and how various road and intersection layouts were failing, safety-wise. At that time there was no road safety auditing of designs in Sweden. Had the Swedes had a history and culture of crash/blackspot investigation, they would have had road safety auditing of designs and they would have stopped using substandard designs like the entry ramp in Figure 5.3. It is vital that road safety engineers gain their experience as widely as possible.

Perhaps examples like this explain why the Swedes, in their Vision Zero, chose to focus on fatal and serious (disabling) injuries: a limited history of

---

66 For example see Ekman, 2014, slide 12.

crash investigation and understanding its benefits for safe road design.[67] It doesn't however, explain why people from countries that do have a history and culture of blackspot investigation (e.g. Australia and the UK) are tripping over themselves to copy the Swedes and make fatal and serious injury crashes the focus of a supposedly 'Safe System'. Actually, I just led you astray in that previous sentence — it explains it completely: the proponents of the Safe System are not the people in these countries who have a depth of understanding of crash investigation and an appreciation of the value of safely engineered road infrastructure. They are the people who think that the way to achieve a safe road system is to take a broad strategy and tell road users how to behave, then put in place strictly enforced controls, supposedly for the road users' own good.[68]

Details vs. population-level strategies

This discussion about the origins of Vision Zero in Sweden is intended to get to the heart of the Safe System.

It is obvious by now that I see success in road safety as a matter of getting to the details. Not only do we need to disaggregate data to establish, for example, which road users are crashing across the network; we need to look at the details of the crash causes at each problem site. Understanding this will let us target the causes with a specific, effective solution. By contrast, proponents of the Safe System propose a population-wide approach like lower speed limits. Case Study 2 in Chapter 2 illustrates the contrast: do we say 'Gosh we've got a lot of crashes. Crash rates are related to travel speeds. So let's lower the speed

---

67   Of course the development of Sweden's '2+1 with barrier' designs came from an appreciation of the crashes that were happening. Let me not diminish that effort. Nonetheless some aspects of those schemes are not as safe as they might have been. Also see the next footnote.

68   Perhaps this also better explains the likely situation in Sweden. The people with an understanding of crash investigation and its benefits for safe road design were probably the ones who in the 1990s were developing the '2+1 with barrier' treatment to cut the high crash rate on those 13 m wide roads. Meanwhile, another group perhaps without such a depth of understanding were elsewhere developing Vision Zero with its focus on fatal and serious injuries.

limit' or do we look at the details of the crashes, identify the causes at each crash problem site and then put in place engineering solutions to stop people making the mistakes that lead to those crashes?[69]

As mentioned earlier, Safe System proponents argue that road safety should be addressed as a public health issue. For example:

> It is well and truly time that public health strategies were brought systematically to bear on the traffic safety problem.[70]

This means applying a population-level strategy whenever a low-level risk is spread widely across a community. Two of the three 'Johnston et al' authors, Dr Ian Johnston and Dr Carlyn Muir are psychologists. Training for psychologists and medical practitioners relies heavily on epidemiological data[71], so their approach is understandable. But I do not think it is helpful.

One might conclude from reading Johnston et al that this epidemiology-based approach of shifting the behaviour of an entire population with a low-level risk is accepted as the only approach in public health. But Harper, 2009 describes the example of the eye disease glaucoma where two-thirds of cases come from those with moderate-to-low levels of intraocular pressure (the cause of the disease), yet fully one-third of cases come from 'the extreme tail of the distribution', just 4% of the population. This being the case, any effective public health program in this area needs to target the *deviant* or *high-risk* individuals and not simply rely on a shift in the whole population.

Harper makes another point that achieving a shift in the whole

---

69   Also see the second half of Chapter 8 for further discussion of specific crash problems and their solutions on the road in Case Study 2.

70   Johnston et al, 2017, p 105.

71   Epidemiology may be defined as the study of disease in populations. It deals with the incidence, distribution and control of diseases in specific populations. Thus it is not interested in the details of disease processes in affected individuals.

population is likely to be effective (in reducing the incidence of a health problem) if the distribution is normal (as with a bell curve). But where the distribution is skewed to the right (more people at the hazardous end) it is equally arguable that targeting the high-risk segment will be just as effective or more so.

Now consider two road safety issues: alcohol and speed.

Harper comments that many years ago and in the absence of solid data, Dr Geoffrey Rose (the widely-esteemed proponent of population-level public health strategies) speculated that a large part of alcohol-related crashes could conceivably be due to the accumulated risks of large numbers of drivers with small amounts of alcohol. Subsequent research showed this was not the case. As Harper puts it:

> Thus while it is true that accident risk increases with any alcohol consumption, it does so exponentially, making the contribution of light drinkers to alcohol-involved crashes minor despite their greater representation in the population.

Thus targeting individuals who are over a reasonable blood alcohol concentration (BAC) limit is more effective than seeking a population-level shift amongst low-risk drivers by reducing the permitted BAC limit (currently 0.05% in Australia).

If we look at speed and collision risk it's a similar situation. This topic is discussed in greater detail in Chapter 7, but for now let me mention the often-referenced speed / crash risk report by Kloeden et al, 1997. That report includes a graph that shows the risk of involvement in a casualty crash increases exponentially with increasing travel speed above the speed limit.[72] The conclusion must be the same: the contribution of drivers travelling a little faster than a reasonable limit is minor, despite their greater representation in the population.

---

72   Kloeden et al, 1997, Figure 4.3 Travelling Speed and the Risk of Involvement in a Casualty Crash, Relative to Travelling at 60 km/h in a 60 km/h Speed Limit Zone.

In other words, the speed enforcement agenda needs to focus on those drivers who are exceeding reasonable limits by a significant amount. It should not be focusing on the vast majority of drivers and regarding them all as potential behaviour modification targets (or potential law-breakers, for that matter) in an effort to achieve a population-level shift in behaviour to new, unreasonably low travel speeds. Mention above about the 4% of the population at high risk of glaucoma who need to be targeted by the medical profession reminds me of a former senior traffic police officer's comment that just 4% of the population causes 90% of policing problems. He understood: the enforcement effort needs to be at that end.

OK, but what about Harper, 2009's comment (above) that occurrences which are normally distributed as a bell curve are amenable to programs that seek a population-wide shift in behaviour? The distribution of free speeds along a road is typically a bell curve, so surely a shift to lower speeds (via lower speed limits or lower enforcement tolerances) should be sought? It is here where the *public health strategy* approach to road safety by Safe System advocates — and the focus on fatal and serious injury crashes — are exposed:

► Just because a strategy *can* be adopted, it doesn't mean it automatically should be — there needs to be a good reason for doing it and the way to determine that is to look at the details. (Just like reducing speed limits around all schools — there needed to be a good reason, but the details showed that there wasn't). Reducing speeds is only of value where the cause of the crashes is speed-related. General 'evidence' that crashes increase with higher travel speeds is meaningless if the details in the particular case show that it's not applicable.

► Looking only at fatal and serious injury crashes is just another way of avoiding details. Looking at just 1% of crashes won't explain much and it almost begs for some magical strategy to be adopted.

► When the details of the *public health strategy* approach are

56

examined, it's evident that it's not as simple as suggested by Johnston et al. In the public health sector an effective strategy is not just about applying population-level strategies to people with a low-level risk. Depending on the details of each health problem, different approaches need to be taken, including targeting those most at risk.

▶ From the discussion above, it is evident that the transferability of the public health approach[73] to speed management is highly questionable.

Whether or not you believe the applicability of the public health strategy approach to road safety, effective solutions are going to result from:

▶ Looking at as big a crash sample as possible, appreciating that crash severities may be wrongly recorded.

▶ Looking at the details in the crash data, to understand causes at each problem site.

▶ Applying effective, targeted solutions.

Which gets us back to where this chapter started: road safety effort should not be limited to the small shaded one percent 'Area of concern' in the sixth column of Figure 5.2; it should be applied to the whole 100% shaded 'Areas of concern' in the right-hand column. If we do that, we are likely to be more effective in reducing the numbers of fatal and serious injuries.

That is one of the principles behind my alternative to the Safe System which I will explain in Chapter 10.

---

73   That is, applying a population-level strategy where a low-level risk is spread widely across the whole community.

Chapter 6

RISK, FREEDOM AND EVERYTHING ELSE

With zero not possible, its pursuit resulting in worse safety (more crashes) and the focus on fatal and serious injury crashes likely to be counterproductive (resulting in more of these severe crashes), the foundations for the Safe System appear shaky. But there are more fundamental issues at stake: issues about how we want our society to operate — issues like risk, freedom, responsibility and trust.

Risk means being exposed to the possibility of an adverse outcome (danger, harm or loss). To consider the size or scale of a risk we multiply the likelihood (probability) by the severity of the outcome, were it to happen. In road safety, the severity of various outcomes is typically well established. The likelihood or probability is harder to establish. Programs like ARRB's Road Safety Risk Manager (McInerney, 2002) provide quantitative advice. Otherwise, risk can be assessed in broad terms, based on broad assessments of severity and likelihood.

Freedom is generally thought of as being able to conduct one's life as one chooses. Within a society we all have desires, responsibilities and obligations; we all impact on others and they on us. So I might define freedom as being able to conduct one's life as one chooses, without imposing unreasonable impacts on others. Every society (and separately, every government) will have a different view of what is reasonable and unreasonable.

While I was typing in 'Sweden's reduced crash rate', Google helpfully offered me 'Sweden's reduced prostitution'. Far more interesting! From Wikipedia[74] I learned that:

▶ In the 20th century Sweden had eight inquiries or commissions that investigated prostitution.

▶ In 1999 a law was passed banning (criminalizing) the purchase of sex, but not its sale (a later amendment banned sale via pimps).

▶ Before the law, Sweden had less prostitution than other European countries.

▶ Opinion polls show high public support for the law.

▶ The law has been called the *Swedish model* and has been adopted or considered elsewhere.

▶ A 2010 inquiry report found the law had halved street prostitution.

That sounds like a different society from the one I live in, in Victoria, Australia. Here the prostitution issue was largely settled in the 1980s when regulated brothels were legalised, after widespread consultation including with people involved in prostitution. There is no right or wrong in this: I might think Sweden has a national obsession about a lesser-order problem, but I don't live or vote there. So, until some do-gooder suggests we adopt the *Swedish model* here, it's not my concern.

The blood alcohol concentration (BAC) limit for drivers in Sweden is 0.02% (down from 0.05% in 1990). In Australia, it is 0.05% ('Point O Five'). In New Zealand and Scotland, it was recently lowered from 0.08% to 0.05%. In the other countries listed in Chapter 4 with low crash rates, it's 0.02% in Norway (down from 0.05% in 2001), 0.05% in Denmark and the Netherlands and 0.08% in the U.K. (except Scotland). Apart from anything else, each of these expresses how a society views the freedom of the individual differently.

---

74 *https://en.wikipedia.org/wiki/Prostitution_in_Sweden* (Dec. 2017).

Adams, 1985 refers to 'a politically powerful temperance tradition in Scandinavia'[75]. It's reported that 76.5% of Swedish drivers support an even lower limit than 0.02%.[76] Ross et al, 1992 advise that until 1990 all drink driving offences in Sweden resulted in jail, then:

> In Sweden, abolition of mandatory jail was accompanied by reducing the limit of tolerated BAC from 0.05% to 0.02%, arguably punishing not just impaired driving but all drinking associated with driving. (However, technical changes in the measurement of BAC have diminished the change in practice.)

These days for a minor drink driving offence 'the person is . . . given a fine', the amount being related to the offender's income.[77] Ross et al comment that this reduction in the severity of Swedish drink driving laws was accompanied by a significant decrease in crash fatalities. Interestingly, prior to the change alcohol was found in the blood of 24-29% of deceased drivers, not much less than the 32% figure for the USA 'which has far more impaired drivers on the road'. The mean BAC of deceased impaired drivers was 0.17% in Sweden, much the same as 0.16% in the USA, leading Ross et al to conclude:

> the deterrable [people who could be readily deterred from drinking and driving] may have been eliminated from the roads, leaving a highly dangerous minority of alcoholics.

My conclusions from this are that Sweden's *tough on drink driving* approach relates strongly to their national culture about alcohol, but the severity of punishment has little impact on levels of alcohol involvement in crashes. We should at least be wary of importing road safety *solutions* from other quite different societies.

Yet in the Safe System era in Australia we now see even harsher penalties like automatic loss of licence and ignition interlocks for anyone *ever*

---

75   Adams, 1985, pp 121-122.

76   Meesmann & Rossi, 2015, p 19.

77   Meesmann & Rossi, 2015, p 20.

caught at a BAC of 0.05% or more. So far as I know the only way for a person to easily tell if they are *over the limit* is for them to be stopped by police at a random breath test. Apart from the increasing severity of penalty appearing to be counterproductive, the whole system has been set up with little accountability and little interest in promoting personal responsibility.

With the Safe System the most obvious issue around risk and freedom is mobility. According to Haight[78]:

> [W]here mobility and safety are in conflict, the conflict relates to speed.
>
> A realistic approach to the problem of reconciling mobility and safety must, to begin with, acknowledge that both are social goods which have value. In very simple terms, we may think of three-way optimisation: maximise safety, maximise mobility, minimise cost. A little reflection will show that any two parts of this triad can be reasonably well satisfied at the expense of the third. The goal of public policy is to optimise all three; this seems to be nearly impossible.

So when Johnston et al say things like 'human life is sacrosanct' and 'the elimination of disabling injury and death [is] nonnegotiable (sic)' and put forward proposals for universal, automatic, minute by minute enforcement of their safety solutions (i.e. lower speed limits), I think of Haight's comment:

> If the transport community treats safety superficially, it is true *a fortiori* [*with even greater evidence*] that the safety community treats transport issues — if at all — with contempt. At best, authors writing about traffic safety brush mobility aside as being of no consequence.

After describing some of the absurd suggestions by others to eliminate crashes, like urban speeds should be 5 km/h and trips less than 5 km should be done walking or by bicycle, Haight adds some absurd

---

78   Haight, 1994, pp 12 & 14.

suggestions of his own including 'cancel licence after first offence', 'drive only in daylight' and 'put median barriers in every street'. He concludes by saying:

> The simple observation that none of these suggestions would be taken seriously is a sufficient indication that we do not in fact consider life to be priceless or time without a value.[79]

The whole concept of zero fatalities and serious injuries being achievable is based on the view that mobility has no value in an industrialised society, that the preservation of life[80] is the only morally acceptable objective and that there is no limit to the amount of money that should be spent to preserve life in the name of road safety. I dips me lid to Professor Claes Tingvall and his associates for managing to convince an entire national parliament that this is a way forward and then go on to convince other road safety 'experts' and nations, as if *The Emperor's New Clothes* is Swedish, rather than Danish.[81]

Miller, 2018 makes the point:

> Ideals often contradict. Safety conflicts with freedom, freedom with equality, equality with opportunity, opportunity with safety. We can promise to live up to each and every ideal, then find ourselves in positions where that's impossible, or at least inadvisable.

---

79   Haight, 1994, p 13.

80   Not all life, just the life of people who willingly participate in the traffic system.

81   As explained by Adams, 1985 'The fable [of the Emperor's new clothes] suggests that once an idea, however preposterous, becomes accepted by, and espoused by, established authorities it can become extremely difficult to dislodge. The idea becomes self-reinforcing. Authorities cite prior authorities, until the idea accumulates an authoritative pedigree. The idea acquires its own defence mechanism. Anyone incapable of seeing the Emperor's new clothes is 'unfit for his situation, or unpardonably stupid.' The fact that large numbers of other people believe the idea can become sufficient reason for believing. After a while evidence is no longer required.'

Is there something in the psyche of road safety 'experts' that prevents them from acknowledging this complexity of life and instead leads them to take a one-dimensional view? Johnston et al appear quite happy for us to go bankrupt, in the pointless pursuit of zero fatal and serious injury crashes via low speed limits:

> It is a simple matter to ensure all speed limits are signed using flashing LED displays rather than static signs.[82]

Figure 6.1
Beaut ways to bankruptcy: require all speed limit signs to be flashing LED

But not only do they appear happy for us to go bankrupt in this one-dimensional world. We are also required to live in a police state as pervasive as Stalin's. Here is their suggestion for enforcing speed limits:

> With technology to register a vehicle's speed continuously, one might even consider an approach analogous to London's congestion charging scheme. Individual vehicles could receive a regular account where they are charged for their time above the speed limit, on an increasing scale if behaviour is repeated.[83]

---

82   Johnston et al, 2017, p 138.

83   Johnston et al, 2017, p 138. This reminds me of that old joke: How do you know that the people advocating lower speed limits as the solution for every road safety problem aren't circumcised?

With the rapid development of intelligence in algorithms used for automated detection and enforcement, automated enforcement will appeal more and more to governments and human intervention and discretion will occur less and less. What will be enforced? Exceeding a parking time limit. Rolling through a Stop sign. Jay-walking. Anything.

Laws will be changed to reduce the accountability of the enforcement agency, if only to cope with the sheer volume of infringement notices. In Victoria we already have laws that specifically prevent human assessment of accuracy or fairness with automated enforcement. In law, the speed limit at a speed camera site is not what any speed limit sign states, nor what the Road Rules may say about the applicability of that sign at that particular time or on that particular day. No, in law, the speed limit is what an authorised 'justice' department functionary says it is (by writing it on an image) after reading it from an authorised list.[84]

Woodrow Hartzog at the Cumberland School of Law at Samford University in Birmingham, Alabama is quoted as saying:

> If individuals know they are likely under surveillance, they have decreased autonomy and are less likely to engage in important kinds of behaviour necessary for human development.[85]

The more that Safe System advocates call for speed limits based on the level of protection[86] rather than on the road, roadside and traffic conditions, the more enforcement will be needed, the more that enforcement will be automated and the more governments will seek to remove or diminish human intervention in the process. No computer code is bug-free, so errors will occur and leave road users having to prove their innocence. More insidiously, less discretion and common sense will be involved, as already evidenced by Victoria's speed camera laws.

---

84 Road Safety Act, Victoria (1986), Sections 81(1) and (2).

85 see Moskvitch, 2013.

86 In other words, based on the potential severity of outcome, without any regard to the likelihood of the collision occurring. See Johnston et al, 2017, p 138.

Under communism, Soviet Russia had an extensive network of secret police, all in the name of the security of the population. Under the Safe System we can look forward to having an extensive network of automated enforcement cameras, all in the name of the safety of the population. It is perverse in the extreme for Johnston, Howard and Muir to be preaching morality and telling us that 'human life is sacrosanct' while proposing ways for pervasive control of our lives. At the end of their Chapter 8 their suggestion to 'mobilise the "silent majority"' is reminiscent of the long-successful Chinese practice of using village informants to weed out miscreants and keep society under control.[87]

All activity involves risk. If you want no risk in your life, stand still — though that may lead to the risk of boredom or obesity instead. If we do not encounter any risks, it is likely we will become increasingly anxious, as encountering risks and dealing with them allows us to keep perspective and gain confidence in our daily dealings with life. The fewer risks we encounter, the more distorted our perception of them. For example, our roads are much safer now than 50 years ago, yet more parents are anxious about letting their children walk unaccompanied to school.

In another example, I've had the experience of a T-intersection onto a four lane divided road (with right turn/U-turn lanes, but no left turn lane) near my home being signalised. Despite there having been no recorded crashes involving right turners, no recollection of glass on the road and the site not meeting warrants for red arrow signals, the right turn and U-turn were fully controlled with green/yellow/red arrows. Subsequent efforts to get the red arrows deleted have been met with road authority comments like 'We can't do that — the road safety people won't let us'. So now risk aversion in road safety extends to avoiding being told 'No' by *road safety people* (read *Safe System people*). Not only do we now have one-dimensional Safe System people pursuing a cause in which mobility has no value, but other professionals in road authorities are not prepared to challenge them. Does someone need to point out that this is how bullies get their way, through the

---

87   Johnston et al, 2017, pp 28 and 116.

silence of others? Meanwhile, back at the intersection the right turners sit there, immobile for no good purpose, facing a red arrow signal when there's no oncoming traffic.

Surely the aim with road safety is not the elimination of risk *per se*, but where hazards cannot or need not be eliminated, we should seek to make road users' perceptions of the risk consistent with the actual levels of risk. That is — have no nasty surprises.

In the 1980s Main Roads Western Australia removed zebra crossings on arterial roads, due to their high pedestrian crash levels. Instead they marked painted medians with refuge islands at intervals. Pedestrian crashes dropped dramatically.[88] To some extent this reduced pedestrians' freedom (mobility) but it clearly left pedestrians with a better perception of the actual risks.

In the UK the priorities between vehicles and pedestrians at intersections are undefined. Compare this with Australia where they are clearly defined. Does the UK have a lower crash rate for pedestrians in these circumstances? In 1988 I established that in New Zealand, where a similar situation to the UK existed at unsignalised intersections, crash rates for pedestrians were about half those in Australia (Morgan, 1988). More freedom in Australia, but also more risk: a case of perceptions of the risk not matching the actual levels of risk.

The vexing question of how we provide better mobility for cyclists without increasing their risk has not been solved anywhere — although in Australia we could do some obvious things like banning car parking on major roads with cyclists, so they are not 'doored'. The freedom of drivers (to park), traders (to rely on on-road parking) and Councils (to avoid providing off-street parking) is at the expense cyclists' freedom.

The trajectory of the Safe System in Australia, with its 'primacy of speed management' is distracting us from dealing with these real,

---

88   Moses, 1989: both pedestrian crashes and vehicle crashes (mainly rear-end collisions) dropped by over 80%.

non-speed-related issues about getting more mobility and less risk for particular groups of road users.[89]

Instead of treating road users with contempt and distrust and needing to micro-manage drivers' every move, perhaps we could start by treating them with respect and trust. This is not an original thought. Over forty years ago in the UK Leeming, 1977 said:

> The remedy [to traffic accident problems] will involve trusting the motorist, and many will react with horror. But we already do it in other ways. He runs the country, as most motorists are in responsible posts. It is argued that a man changes his being at the wheel of a car, but that is nonsense. We have in fact trusted drivers in many ways, and saved accidents in doing so. We have relaxed some 30 mile/h speed limits to 40 and saved fatal and serious accidents. We abolished the 20 mile/h limit in 1930, and deaths fell by about 600 the following year. The mini-roundabout involves trusting drivers and works well, and we have relaxed Halt signs to Give Way, without any increase in accidents. All point the same way, that we can trust him.

Regarding the latter point, Stop sign warrants were changed in Victoria in 1990 and nationally[90] in 1994, based on UK experience and were expected to reduce Stop signs from 90% of all control signs to fewer than 5% (the rest being Give Way). Twenty-five years later few regions have brought their signing into line with AS 1742.2 (the 1994 edition onwards) and the overuse of Stop signs remains widespread. Surveys I have done[91] where the change did happen, show that where the network is 95%+ Give Way signs, more people stop at the Give Way signs than ever stopped at the Stop signs when Stop was the predominant

---

89   See Chapter 8 for a discussion of Glen Huntly Road, Elsternwick (Melbourne) where a new Austroads guide ignores the parking / cycling issue while proposing Safe System solutions for that road. Before turning to it, you may like to guess what their main 'solutions' are.

90   Nationally in AS 1742.2, the relevant part of the Manual of uniform traffic control devices.

91   Morgan, 1994, plus additional surveys in 2011 and 2012.

control. The overall number who 'effectively stop' also increased and the number who slowly cruise through, instead of stopping, halved. I have concluded that limiting official intervention to the few really hazardous locations improves safety. Elsewhere, it is important to tell drivers where critical decision points are, but once that is done, giving them the responsibility to make their own decisions — rather than telling them how to act — results in more motorists using appropriate caution. In other words, trusting them improves safety.

Figure 6.2
Trusting drivers improves safety

So in Victoria, we have the irresponsible situation where numerous municipal councils have not, in over 25 years, made one of the most simple, cheap and fundamental traffic safety changes they were obliged to and the state road authority has made no effort to ensure the change happens. Yet VicRoads is now by and large acquiescing uncritically to council requests for lower speed limits. Can't do the fundamentals, but are keen to play with the Safe System. When in 2015 I wrote at length to VicRoads requesting they address this 25 years of inaction (and included my survey results and conclusions), their inane response inferred that maybe the warrants in AS 1742.2 were wrong and we should allow more Stop signs. It was a reminder to me that the micro-management school of road safety is alive and well.[92] Again, they can't do the fundamentals but are keen to play with the Safe System.

Little by little our freedom — as well as our safety when we use our roads — is being eroded by the new breed of Safe System state functionary.

---

92 It also reminded me of the famous paper by Kruger and Dunning in 1999 about the problems unskilled people have recognising their own incompetence: '... people who are unskilled in these domains suffer a dual burden: Not only do these people reach erroneous conclusions and make unfortunate choices, but their incompetence robs them of the metacognitive ability to realize it.'

Chapter 7

REGIME DESTRUCTION

In Chapter 1 I listed the elements of what I call the *Safe System Manifesto*. They include:

► Speed limits must match the level of protection provided.

► If they don't, speed limits must be lowered.

If there is one belief the public (when not themselves driving) has resolutely held onto in traffic matters it is that you can control speeds simply by installing a speed limit sign. Traffic going too fast? Demand a lower speed limit. Never mind that the evidence indicates the speed limit is about the last thing that dictates travel speeds. A UK study (Silcock et al, 2000) reported that:

> It is clear from our surveys that drivers generally make their own assessment of the speed at which they will drive, irrespective of the speed limit, based on their own judgement of the road environment. As well as physical dimensions and layout of the road, this includes prevailing traffic conditions, and whether a road is perceived as urban or rural.

> Dual carriageways with 30 or 40 mile/h [48 or 64 km/h] limits were particularly susceptible to speeding.

Even in Victoria where drivers these days are more fearful of being caught by a mobile speed camera than in most other jurisdictions in the world, surveys on roads with speed limits that are low for the conditions regularly show travel speeds well above the sign-posted limit.

Yet with the Safe System, we now have a whole road safety strategy based on the idea that if all else fails (or quite possibly, if all else might be effective but costs money to build) the solution is simply to reduce the speed limit to achieve better safety. And we have a whole road safety 'industry' happy to go along with this, rather than confront the reality. What's happening?

In Chapter 1 I suggested that the second dot point above could be re-phrased as meaning 'the Safe System requires the destruction of the old speed limit regime'.

While I was a local government traffic engineer through the 1980s, Victoria had 75 km/h speed zones, which were a leftover from the metrication of 45 mph in 1974. Across Australia the default speed limits were 60 km/h (urban) and 100 km/h (rural). The speed limits used in Victoria were 60, 75, 90, 100 and 110 km/h. Several traffic engineering people in the state's road and traffic authorities and a few of us in local government thought the speed limit in local streets should be 50 km/h and that speed limits above that should be in increments of 10 km/h, as they were elsewhere in Australia. The speed regime we thought should be in place is shown in Table 7.1.

And experience has shown that this was pretty well right — perhaps apart from 110 km/h being too low on the highest standard rural freeways with moderate traffic volumes. For the form and layout of typical roads in Australia, these listed values approximate the speed most drivers will adopt in most situations when 'driving to the conditions'[93]. Unlike what many in the micro-management school of road safety may believe, this is not a precise science and there is nothing inherently perfect about exactly 50 or 60 or 100 km/h that requires exact enforcement.

93 These listed speeds (except perhaps the 90 km/h) are likely to approximate the 85th percentile speeds on typical roads of these types. See later in this chapter for an explanation of '85th percentile speed'.

Table 7.1
Speed limits sought in the 1980s

10 km/h	Shared zones (equal pedestrian/vehicle rights).
40 km/h	Specific high pedestrian locations or school crossing approaches, where there were speed-related problems, applied on a needs basis (not blanket/as-of-right at all schools or all strip shopping centres).
50 km/h	Local streets. General urban limit.
60 km/h	Undivided urban arterial roads ('arterial' being a functional classification before it was later erroneously commandeered by VicRoads as an administrative term to mean a state road).
70 km/h	Divided urban arterial roads with driveway access.
80 km/h	Divided urban arterial roads with no or limited access (e.g. ones with service roads). Low standard urban freeways. Minor rural hamlets.
90 km/h	Urban fringe. Problem locations where taking the edge off 100 km/h speeds would reduce specific crash problems.
100 km/h	Urban (busy) freeways. Rural highways. General rural limit.
110 km/h	Rural freeways well away from Melbourne.

After a short period of sensible speed limits approximating the list in Table 7.1 from ~1990 (in Victoria, at least[94]), things started to go plonkingly wrong when, as a knee-jerk political reaction to some high-profile crashes, the 'responsible' NSW Minister decided to apply 40 km/h limits around all NSW schools. Pressure to copy this drifted across the border. As a result, instead of 40 km/h — or 60 km/h on high speed roads — being applied in Victoria (at the start and end of the school day) on a needs basis at children's crossings with a problem,

---

94   Thanks to people like Ted Barton, John Cunningham and Ken Ogden.

these limits were soon enough applied around all Victorian schools, whether there was a problem or not.

The disconnect between solution and problem became official VicRoads policy — as it had in NSW. And with the focus of the Australian version of the Safe System being 'the primacy of speed management', practitioners were repeatedly told and quickly believed that any reduction in speed limits must be good for safety. Never mind that:

▶ The 1996 MUARC study (discussed in Chapter 2) showed that it wasn't quite that simple and wasn't always true, and

▶ Quayle, 1999 describes several studies where unrealistically low speed limits were raised and the speeds either stayed the same or fell, e.g. in Perth an increase in the limit from 60 to 80 km/h resulted in 85th percentile speeds falling from 87.5 to 82 km/h. 'Similar results flowed from an increase from 60 to 70'. Other studies on the Eyre Highway (W.A.) and in California and the Netherlands showed speeds dropped or did not change when speed limits were raised. [95]

It's back to where this chapter started: a speed limit does not automatically equal the travel speed.

As Quayle comments, the language around speed has been 'mangled'[96] so that a travel speed a little above the signed or default limit is called 'low-level speeding', thereby associating those speeds with what we typically regard as 'speeding': travelling at a speed that is inappropriate for the conditions. To me it's more like manipulation than mangling: seeking to achieve guilt by association.

---

95   And as if to prove the point, Sweden's Vision Zero has led to 100 km/h undivided roads being limited to 80 km/h, contributing to only 44% of drivers on Sweden's national roads keeping within the speed limit (see Chapter 6).

96   Quayle, 2015, p 237. See examples in Johnston et al, 2017, p 130.

Probably the most quoted study worldwide about the risks of travelling at a higher speed is the 1997 study by Kloeden et al based on in-depth crash investigations and other surveys in Adelaide. The principal conclusion was:

> In a 60 km/h speed limit area, the risk of involvement in a casualty crash doubles with each 5 km/h increase in travelling speed above 60 km/h.

A further conclusion was:

> Even travelling at 5 km/h above the 60 km/h limit increases the risk of crash involvement as much as driving with a blood alcohol concentration of 0.05.

The first point to be made is that the generality of the first conclusion supports the findings made by Newstead and Mullan, 1996 and referred to in Chapter 2: there is a greater crash risk where a speed limit (because it is artificially low) leads to greater variations in travel speeds along a road.

Despite Kloeden et al, 1997 being widely quoted around the world by people keen to see lower speed limits, more enforcement or lower tolerance levels for enforcement, the study also has its critics.

Quayle, 2015 is highly sceptical of the transferability of this study's data and conclusions (about speed/collision risk), due to Adelaide's unique condition of being a large city without urban freeways and with a largely grid-iron pattern of streets, leading to unique speeding, rat-running and barging-in behaviours. He further makes the point that South Australian fatality rates are consistently higher than other jurisdictions due to the absence of (safer) freeways in Adelaide.

Lambert, 2002 in his review states:

Further analysis shows that [the] Kloeden et al (1997) report must now be considered to be seriously flawed [and] that the analysis:

► Does not support that there are any magical properties of the number 60 as in 60 km/h

► Fails to highlight that outcomes only apply to free travelling speed crashes (about 28% of serious crashes)

► Fails to recognise that a high BAC applies to the whole trip whereas free travel speed applies to only part

► Fails to adjust for the impact of black spots and black links on the findings, and

► Fails to recognise that speed enforcement does not take place at the crash sites included in the study [A reference to the 1997 study's recommendations about speed enforcement].

If I understand Lambert correctly, he is saying that while most crashes occur at or near major intersections, most speed enforcement occurs mid-block where speeds are higher and where those higher speeds are not associated with higher risk. There is no evidence that drivers who exceed the speed limit mid-block do the same through major intersections. In other words, a low enforcement tolerance at mid-block locations is catching plenty of drivers but is of little value in reducing crashes. The sites in the 1997 study were predominantly blackspots. Seventy-five percent of the crashes occurred at intersections, although only 15% of travel happens in the vicinity of intersections. Thus the risk calculations can't be extended to the whole network. Finally, there is nothing magic about '60 km/h'. If the speed limit is, say, 70 km/h, it will be the scale of speed variation away from 70 km/h (not 60 km/h) that will indicate the level of risk. And if all drivers saw 70 km/h as 'driving to the conditions' and they all travelled at that speed, there would be no increased risk.

The simplistic (and wrong) conclusion by many who have seen Kloeden et al's primary conclusion is that if everyone reduced their speed by 5 km/h the number of crashes would halve (because +5 km/h doubles the risk and risk leads to crashes). If that's the case then Victoria's TAC, which led that state's intense 'Wipe Off 5' campaign in conjunction with a reduction in enforcement tolerances isn't letting on about it. Amongst copious documentation about the 'success' of their Wipe Off 5 campaign — and reference to Kloeden et al's primary conclusion — there is no information at all on their website about reductions in crashes that actually occurred.

Bobevski et al, 2007, reviewing Victorian road safety initiatives associated with the Wipe Off 5 campaign in 2000 to 2002 advise that the following initiatives had the following results:

▶ Increased mobile speed camera hours (from 4,200 to 6,000 hours) resulted in a 3.25% reduction in monthly casualty crashes (though not statistically significant) and a 43% reduction in the risk of a fatal outcome in a crash.

▶ Use of flashless mobile speed cameras (instead of ones that flash) resulted in a 2.1% reduction in monthly casualty crashes, but an increased risk of a fatal outcome in a crash.

▶ The reduction in enforcement tolerances[97] resulted in no significant change in monthly casualty crashes. No mention is made of any effect on the risk of a fatal outcome in a crash, which I take to mean that there was no effect.

▶ The reduction in the default urban speed limit (generally applicable in all local streets) from 60 km/h to 50 km/h in 2001 resulted in a 13.5% reduction in casualty crashes.

---

97 While not advised in the report, it is my understanding from high level advice at the time that the enforcement threshold reduction was 4 km/h, from being booked at 10 km/h over, down to being booked at 6 km/h over the limit.

When I and others were seeking to get the default urban speed limit (for local streets) reduced to 50 km/h in the 1980s and 1990s, we regarded it as an act of faith (a gut instinct) that it would result in lower speeds and a reduction in crashes. There was no clear evidence it would be effective. As it turns out, it was the one significantly effective action of the time.

Given that drivers in Victoria reduced their speeds by around 5 km/h in response to the Wipe Off 5 campaign[98] but that this resulted in no significant change in monthly casualty crashes, then Kloeden et al's finding about an increase of 5 km/h doubling the risk of a crash clearly does not mean casualty crashes are halved or doubled with a 5 km/h change in travel speeds. Nil would be a closer estimate.

The logical conclusion about Kloeden et al's work is that it's a case of *correlation does not mean causation*: while the people who were involved in more crashes were ones who drove faster, we cannot necessarily conclude that the one was caused by the other.

If a micro-study was able to establish that those people who drive over the limit are more prone to inattention or risk taking, then there may be a case. But it may simply be that those who drive over the limit and who are also prone to inattention or risk taking are the ones who have more crashes, while others drive over the limit but do pay attention and don't take risks and are not involved in a greater rate of crashes. The evidence suggests this is the case.

In each trip on the road, or even in a decade's trips on the road, the chances of being involved in a casualty crash are minute. A doubling of that risk is still 'minute'.

Or as Johnston et al put it:

---

98   Bobevski et al, 2007, Figures 7.1 and 7.2: mean speeds and 85th percentile speeds in 60, 70 and 80 km/h zones reduced by amounts very close to 5 km/h.

The likelihood of a casualty crash on any individual trip for any individual driver is extremely small, and while driving a small amount over the speed limit may double that risk, we need to acknowledge that *doubling an extremely small risk still leaves a very small risk*. So, *at the level of an individual driver on any individual trip*, the daily experience of speeding may be technically criminal, but it is not demonstrably dangerous or immoral.[99]

All of this simply exposes the efforts of Victoria's TAC and others who try changing road user behaviour directly in the name of road safety to be a waste of time and money. Our experience on the road is overwhelmingly that what we do on a daily basis is safe, so long as we watch out for the occasional risky or inattentive person.

The attitude of the road safety 'experts' in Australia is the opposite of trusting people. I recall driving in the UK in the 1980s where, on 70 mph (110 km/h) motorways the left lane ran at 70 mph, the centre lane at 80 mph and the right lane at 90 mph. I'd return to Victoria, where adherence to speed limits was always much better, and hear police and road safety people berate us for our speeding and poor driving. No one ever thanked us for the generally very good job we did, driving close to the speed limit. It was quite evident that in Victoria any attitude problem was not primarily with road users. Even in 2017 my experience driving in the UK was that speeding (i.e. going 10 mph+ faster than the limit) is still widespread, in ways not evident in Australia. And yet the UK has a fatality rate per 100,000 population that is half that of Victoria. Disdain for and distrust of motorists needs to stop. The Safe System is supposedly based on accepting and working with the limitations of road users, yet there is this colossal blind spot with Safe System advocates who don't want to work with human behaviour as it is. They want to change it because they know what's best for people.

---

99   Johnston et al, 2017, p 130. Note the manipulation of English: 'a small amount over the speed limit' becomes 'speeding'. Also it's not 'technically criminal', it's 'technically illegal'; this is not something criminal law involves itself with.

My alternative to the Safe System in Chapter 10 seeks to avoid this approach.

Here's my view:

► We need to accept and work with the people we have (i.e. the road users). Most people are reasonable and we need to trust them. A small proportion is not reasonable and we also need to acknowledge that and deal with them, but we should not view the majority as untrustworthy.

► Most people's daily, monthly and yearly experience is that they survive their journey without any great scares. Build on this. Most people have a reasonable sense of risk in many situations. But there are definitely some parts of the traffic system where the actual risk is greater than the perceived risk. This is something experienced practitioners do know. But instead of trying to change road users' behaviour to reduce the risk in these locations, we need to alert them to the actual risk level (so they can take action) or we need to re-engineer the traffic space to reduce/ remove the risk (also see Figure 2.1). In either case we need to match their perception of risk with the actual level of risk.

► We need to stop crying wolf, for example by implementing low speed limits where they are not needed, such as around *every* school or where Councils ask for them on the basis of 'If they've got one, we want one too'. This hides the actually risky locations: because drivers aren't stupid — they know most of our low speed limits are pointless, like most of the unwarranted, over-used Stop signs. But which ones? Over-use increases the risks at actually risky sites.

► As a profession, we need to be a little more discerning than simply accepting whatever piece of dogma is in favour at the time, like deciding that mobility doesn't matter and zero is possible. We need to trust road users and use the actual evidence that has been gathered through research and experience to work with road users to help them make effective decisions, not micro-manage their every move.

Figure 7.1
Crying wolf: a 40 km/h school speed zone on a 6-lane 70 km/h divided arterial road. The school is on the far side, fronting a completely sepa-rate service road. Crossing supervisors are on duty at the signalised intersection. All turning movements are fully controlled, except one left turn slip lane with a zebra crossing. How could this arrangement — without the 40 km/h limit — be described as 'unsafe'?

Which brings me to ...

85th Percentile Speeds

85th percentile speeds[100] used to be the most significant (but not the only) factor in the determination of the speed limit along a section of road. The 1986 edition of AS 1742.4 stated:

---

100 The speed at or below which 85% of vehicles are observed to travel under free-flowing conditions past a nominated point. The 85th percentile speed is approximately 'one standard deviation' above the mean (~average) speed.

81

The [85th percentile] speed is representative of the general perception of a reasonable travel speed on a particular section of road. Although other factors may indicate that this perception may need to be modified by the imposition of a lower limit, such a limit may be progressively ignored to an increasing extent by drivers as the difference between the 85th percentile speed and the posted speed limit increases. [101]

These sentiments are very much out of favour these days. That edition of AS 1742.4 further stated at clause 7.2.1:

In determination of speed zones, the following criteria should be considered for the particular length of road:

(a)   85th percentile speed

(b)   Roadside development

(c)   Road characteristics

(d)   Traffic characteristics (including accidents)

By 1999 AS 1742.4 had expanded to include what I describe as reasonably detailed, sensible advice. Table 2.1 'Hierarchy of Speed Limits — Urban Roads' and Table 2.2 'Hierarchy of Speed Limits — Rural Traffic Routes' provided advice very similar to the list in Table 7.1 for the initial selection of a speed limit. The elements to be considered were (in order):

▶ Road function

▶ Existing traffic speeds

▶ Speed environment ('the elements of the road and traffic

---

101 Standards Australia, 1986, Cl 7.2.2.

environment which collectively influence a driver's perception of an appropriate maximum travel speed')

▶ Road crash history.

The latter element included the following advice[102]:

> A poor crash history will often indicate the need for counter-measures other than changes to the speed limit.

If only the state road authority engineers involved in Case Study 2 in Chapter 2 had considered advice like this, that divided road might have had effective countermeasures implemented to tackle the specific crash problems and it would now be safer.

But the ill wind of the Safe System was about to blow over Australia and such advice was to disappear. Vision Zero in Sweden was endorsed in 1997 and the Safe System was adopted in Victoria in 2004. The next edition of AS 1742.4, in 2008, saw the former Tables 2.1 and 2.2 replaced with a table advising that urban arterial roads should have a speed limit of 60 km/h and this might be increased to 70 or 80 km/h where adjacent land was only 'partially built-up' or 'sparsely built up'. Although 85th percentile speeds were mentioned once, that and other previously conventional factors in determining speed limits were buried as 'adjustments' and the appendix detailing measurement of 85th percentile speeds was deleted.[103] The writing was on the wall for 85th percentile speeds; regime destruction was at hand.

In the same year, Austroads' first speed limit guidelines were published (Austroads, 2008). With Austroads representing only member road authorities plus the Australian Local Government Association, the Safe System was upfront and obvious. After a rush of blood to the head in its Introduction in which we are told 'Safe and efficient travel is the cornerstone to a healthy and prosperous society' (I thought it was

---

102 Standards Australia, 1999, Cl 2.3.2.
103 Standards Australia, 2008, Cl 2.3.

something like freedom of speech or democracy) the Austroads Road Safety Taskforce advises us:

> Effective speed management needs appropriate infrastructure, accompanied by education and enforcement to maximise compliance and appropriate travel speeds. It is important that the community is made aware of the changes and the associated benefits. It is critical that police enforce all speed limits.

Goodbye sensible speed limits; we need to be educated, compliant and made aware. But most of all we need to be policed.

The report advises 'Speed limits on much of Australia and New Zealand's road network are higher than the limits many OECD countries set on comparable roads', but then includes appendices that show the opposite is also true.[104]

We are further advised:

> The application of appropriate speed limits forms an integral part of the Safe System approach to road safety .. *and* .. The Safe System focuses on harm minimisation as a philosophy in setting speed limits. Moderation of speeds chosen by drivers and riders is critical in establishing a safer road system.[105]

This is not entirely honest because 'harm minimisation' requires a lot more than a mere 'moderation' of speeds. As explained in Section 1.3 of the Austroads guide, if you want to minimise harm (i.e. stop the following crash types resulting in death or serious injury), travel speeds need to be restricted so that impact speeds are no more than the following:

► Pedestrian struck by vehicle: 20 to 30 km/h

104 Austroads, 2008, Appendices A and B.
105 Austroads, 2008, Section 1.2.

84

► Motorcyclist struck by vehicle (or falling off): 20 to 30 km/h

► Side impact vehicle striking a pole or tree: 30 to 40 km/h

► Side impact vehicle to vehicle crash: 50 km/h

► Head-on vehicle to vehicle (equal mass) crash: 70 km/h

Previously speed limits were required to have credibility, to help with acceptance and enforcement. Safe System advocates set about making credible a *sermo non grata* (unacceptable word).

Vision Zero may have had an immaculate conception in Sweden but for its bastard child, the Australian Safe System to have any legitimacy it was important that the accepted legitimacy of 85th percentile speeds was completely demolished. Through its report the Austroads Road Safety Taskforce set about doing that:

> A traditional consideration in assessing or reviewing speed limits is the determination of the 85th percentile speed of the road. The use of 85th percentile speed has been discontinued by many road authorities as a key factor in speed limit setting and is not supported by the Safe System approach to road safety.

> Some existing guidelines specify that [the 85th percentile speed] is one of a number of factors that should be considered when setting a speed limit. However, setting the speed limit based on unconstrained speed choices is unlikely to deliver an optimum balance between costs and benefits, either for individual drivers, or the community as a whole. It is for this reason that the Australian National Road Safety Action Plan states that moderation of speeds chosen by drivers is critical in establishing a safer road system.[106]

The report then provides a three-page appendix to kill off 85th percentile speeds. This includes:

---

106 Austroads, 2008, Sections 4.2 and 5.

It is true that benefits from speed limit reductions may be very limited if enforcement and public education efforts are minimal. It is also true that actual speed reductions have typically been less than the nominal reduction in speed limit. However, substantial benefits have been observed even when enforcement was not very rigorous and initial speed compliance was poor (by contemporary Australian standards).

So which is it? 'very limited' or 'substantial' benefits? Considering that enforcement of lower limits is going to be increasingly automated and with minimal enforcement tolerances, these comments border on deceit.

We know that as a general relationship lower speeds are associated with fewer and less severe crashes. That's not the point of it. The point is that a lower speed limit is no guarantee of lower travel speeds and thus lower crash occurrences and severities: there is no automatic improvement in safety from installing a lower speed limit. Zero is not possible (see Chapter 2) and if we want more effective and lasting safety improvements, the speed limit system needs credibility.[107]

The report's appendix concludes with an attempt to discredit the evidence that wider speed variations (which typically occur when speed limits are set below 85th percentile speeds) are associated with higher risks and higher crash rates. See Case Study 2 in Chapter 2 for a discussion of actual experience with this issue. That evidence is conveniently absent in the Austroads report. The report paints the speed variation ('speed dispersion') issue as 'controversial' and finishes by stating:

the more logical solution, with much stronger research backing, is to reduce limits and use enforcement backed by public education

---

107 Again I refer to Chapter 6 where it is advised that Sweden's Vision Zero has led to 100 km/h undivided roads being limited to 80 km/h, contributing to only 44% of drivers on national roads keeping within the speed limit. That is a speed limit system that lacks credibility.

to reduce the speeds of the fastest vehicles. This will reduce speed variance, mean speeds, and crash risk.

Welcome to the Safe System police state.

Two clearly different points of view now exist:

▶ The Safe System view that speed limits should be based on preventing death or serious injury, should a collision occur (e.g. if you run off the road or a pedestrian steps out). This is now almost universally accepted within road safety circles and at government policy level — as all jurisdictions' policies endorse the Safe System approach. This is discussed shortly.

▶ The view that speed limits should be self-evident from the road, roadside and traffic characteristics. Drivers should 'drive to the conditions'. Where the actual risks in the conditions are not self-evident, alert the road users or re-engineer the road. This is basically the '85th percentile speeds' approach and many practitioners at the coalface still believe this is the right approach.

Taking 85th percentile speeds as the starting point for speed limits involves a level of trust in drivers. As Quayle[108] puts it:

The 85th percentile is no more than a statistically convenient way of recognising that 9 out of 10 people do the right thing.

Quayle also comments:

Ron Cumming [the person behind Figure 3.1] pointed out long ago that driving is a self-paced task and presumably drivers will choose a speed at which they feel comfortable. ... It has been known for decades that speed choice is influenced by the rate of flow of information across the retina of the eye — the further away the stimulus, the higher the speed choice.

---

108 Quayle, 2015, pp 231 and 232.

This points to how engineers can influence speeds by modifying the layout or appearance of roads, as well as pointing to the futility of trying to manage all speeds with low speed limits. Yet Johnston et al, 2017 also dismiss 85th percentile speeds[109]:

> ... when set at the 85th percentile, the [speed] limit was simply a surrogate for the level of apparently acceptable risk to everyday drivers.

Perhaps the people who dismiss 85th percentile speeds or seek to demolish their legitimacy don't appreciate that by doing so they are also criticising the 'self-explaining road' concept, which is a part of the Netherlands' *Sustainable Safety* initiative. That initiative is typically praised in the same breath as Sweden's Vision Zero. As mentioned in Chapter 3, self-explaining roads match the layout to the speed that is desired by the road authority: if you want people to drive at 80 km/h (or 30 km/h) then design the road and construct features within it so that drivers adopt that speed. That sounds to me like using 85th percentile speeds in reverse.

In Victoria in 2014, we saw attempts by the micro-management brigade to achieve speed limit reductions by stealth, by trying to eliminate 70 km/h and 90 km/h speed limits. This was dressed up as a way of 'simplifying' the speed limit system, though drivers seemed to have no problem with the supposed complexity of 10 km/h increments — at least until all the unwarranted 40 km/h limits appeared around every school. The Minister put a stop to it, thank goodness.

Meanwhile in NSW 85th percentile speeds are not used at all in setting speed limits[110] and since 2011 limits of 70 or 90 km/h are only used by

---

109 Johnston et al, 2017, p 128.

110 Mooren et al, 2011, p 9.

exception.[111] One of the factors that influenced the Victorian Minister was experience with the replacement of 75 km/h limits with 70 and 80 km/h limits in the early 1990s. Before that change (described in Chapter 2), six-lane urban divided roads had 60 km/h speed limits and up to 28% of drivers were being caught by speed cameras (at 10 km/h or more above the limit). Once these roads were converted to 70 km/h, speeds did not increase, but speed camera offences reduced to normal levels. If 15 km/h increments in speed limits were unworkable then, how are 20 km/h increments (60-80-100) going to work now?

Here is my logic challenge to the micro-management brigade. Consider the following:

▶ If nine out of ten drivers adopt a speed based on roadside development, road characteristics and levels of activity on the road and at the roadside (i.e. the 85th percentile speed), this behaviour must be regarded as 'normal'.

▶ Lower limits require more enforcement resources. However, lower limits (or reduced enforcement tolerances which equate to a practical reduction in the limit) do not automatically result in reduced numbers of crashes, as described earlier. So a lot of that enforcement effort is wasted.

▶ Experience shows that the implementation of lower limits results in the actual causes of the crashes not being investigated and not being fixed (see Case Studies 1 and 2 in Chapter 2).

Given these facts, can someone explain why 85th percentile speeds have been banished — other than because the micro-management brigade needs to demonise this foundation of speed management in

111 Speeds on divided roads in NSW appear to be 10 km/h lower than in the list in Table 7.1. Unlike Victoria, NSW has no strong history of access control. There is far less use of service roads. Even the main freeway between Sydney and Melbourne has direct farm accesses within NSW.

order to bring in their new regime, in the pointless quest for zero fatal and serious crashes through lower speed limits under the Safe System.

Not only is the removal of 85th percentile speeds illogical, so too is the replacement Safe System regime. Johnston et al put forward the Safe System view that speed limits must match 'the level of protection offered by the road infrastructure'.[112] If a pedestrian can walk out in front of a car, select a speed limit that avoids the pedestrian being killed or seriously injured *if* that were to happen, regardless of the likelihood. Risk is the combination of the severity of a crash if it happens and the likelihood of it happening. The Safe System approach looks at only half the equation — it does not concern itself with likelihood.[113] There are plenty of features we *could* run into on or beside the road, but what is the likelihood? The Safe System premise that safety is everything (we must achieve zero fatalities and serious injuries) inevitably leads to this illogicality: mobility has no value and crash likelihood is not a consideration. After all, it *could* happen.[114] Their solution to managing risk is to ignore half the risk equation (likelihood) and manage the severity on the basis that the crash will happen. I think it takes a fairly distorted view of humanity and a messianic view of one's own understanding of life to put forward the Safe System approach to speed management.

Obviously speed management is a crucial part of the safe management of our roads. I just happen to think that a little more respect for people will achieve better results.

The speed limits listed in Table 7.1 worked very effectively when they were given the chance, during the 1990s in Victoria (and with 50 km/h

---

112 Johnston et al, 2017, p 82.

113 For example see Lydon & Turner, 2017: 'Speed limits and travel speeds need to be matched to the type of crash possible [rather than *likely*], the protection afforded by the infrastructure and the types of road users using the road.'

114 Equally I *could* win a million dollars in a lottery, as I have a ticket each week. It *could* change my life. But the likelihood is extremely low and I haven't arranged my life on the basis that I *will* win.

finally being adopted as the default urban limit in 2001). The listed values approximate likely 85th percentile speeds in typical conditions and they should form the basis for speed limits across Australia. By contrast, the only 'benefit' of the Safe System approach to speed management is that it paves the way for the wholesale proliferation of automated speed cameras, as urged by the Safe System Manifesto.

Chapter 8

ALL ABOARD — BUT NOT REALLY

Every Australian jurisdiction — state, territory and national — is on board with the Safe System. It's the basis for every road safety strategy. The stated aim of all these strategies is the elimination of fatal and serious injuries on our roads.

Figure 1.1 in Chapter 1 shows the Safe System framework. That is followed by a listing of the essential elements of the Safe System.

As Mooren et al, 2011 explain:

> Under the *Safe System* approach, road and vehicle designers and managers are responsible for designing, producing and managing road travel infrastructure and equipment. Beyond this, road and transport authorities are responsible for putting in place rules and guidance to safe system use — and road users themselves are responsible for abiding by the rules and being alert to injury risks.

There is a simplicity in this framework which potentially aids its understanding. But experience since 2004 has shown that it also allows the framework to be interpreted in countless ways: it can mean what people want it to mean, to justify a range of agendas. Mooren et al, 2011 note that under the Safe System banner NSW has focused on re-fitting its road infrastructure while Victoria has focused on speed compliance.

As a result, the Safe System is something of a moving target. For example:

▶ The original 2004 version omitted 'Emergency medical treatment'. It has since been added.

- ► 'Legislation & Enforcement of Road Rules' has become 'Education & Enforcement of Road Rules' or just 'Enforcement of Road Rules' in some versions.

- ► By 2008 authors at MUARC had reworked 'being forgiving (or accepting) of human error' into 'the limits of human performance[115]' although that appears not to include an appreciation of issues like those in Figure 3.1.

- ► In the 2008 national road safety strategy document, all the 'Safer' items became 'Safe', meaning that the national strategy is that we will have 'Safe Vehicles', 'Safe Roads' and 'Safe speeds', leading to 'Safe Travel' — a rather courageous prediction.

As recommended by MUARC, the Western Australian Road Safety Strategy 2008-2020[116] lists what the state will do under the heading of 'Safe Speeds':

Enhanced enforcement, and

Specific speed limit adjustments to match geographic priorities.

Meanwhile, Sunshine Coast Council's Road Safety Plan[117] lists the following actions under 'Speed management measures':

A19 — Design new residential streets for low speeds

A20 — Traffic calming in existing residential streets

A21 — Identify high risk locations and develop a priority list for possible speed management measures for inclusion in the Ten Year Capital Works Program

115 Corben et al, 2008, p 2.
116 Office of Road Safety, Western Australia, 2009, p 27.
117 Sunshine Coast Council, 2016, p 20.

A22 — Trial new technologies to see what makes a difference

So while one local government authority has worked out that effective speed management has more to do with designing and re-engineering the road space than forcing behaviour change, Australia's most-noted road safety research establishment and the government of Australia's biggest state can only think of speed management in terms of speed limits and enforcement — each under the banner of the Safe System.

Our next-biggest state, Queensland is not actually on board at all. Despite its road safety strategy being based on the Safe System, including achieving zero fatal and serious injuries, the reality is that Queensland's Transport Operation (Road Use Management) Act 1995 states:

Achieving an appropriate balance between safety and cost

(1) Although it may be possible to regulate to achieve the highest level of safety, doing so would ignore the impact of the regulation on the effectiveness and efficiency of road use.

(2) Therefore, this Act acknowledges the need to achieve an appropriate balance between safety, and the costs that regulation imposes on road users and the community.[118]

I happen to think that sounds like a reasonable approach, but considering the discussion in Chapters 3 and 6 it is not the Safe System approach.

Then there are numerous examples of 'We're not quite on board but we need you to believe we are'. Strategies with names like *Towards Zero*, *The Road Towards Zero* and *Working Towards Vision Zero* sound

---

118 Queensland Transport Operation (Road Use Management) Act 1995, Chapter 1, Section 4. Chapter 2, Part 4 also includes some enlightened objectives for enforcement.

as convincing as 'towards making my bed' or 'working towards washing the dishes'.

It's all smoke and mirrors. When Victoria's Transport Accident Commission (TAC) launched its Towards Zero strategy in 2015, Melburnians had the chance to drive along freeways under large banners that proclaimed 'Zero is possible.'. The previous time I had an experience of that kind was when I passed through the gates at the former Nazi concentration camp at Auschwitz, under the banner proclaiming 'Arbeit Macht Frei' (Work Sets You Free).

Figure 8.1
Our micro-management of your behaviour sets you free

Amongst all this confusion and distraction Austroads Research Report AP-R514-16 *Achieving Safe System Speeds on Urban Arterial Roads: Compendium of Good Practice* is a refreshing beacon of common sense, full of useful engineering-based treatments to help manage speeds on arterial roads. Suitably implemented, these are the types of treatments that can produce effective ongoing speed management. But much of it is not actually Safe System. As the Introduction explains:

> The treatments that produce Safe System speeds are most desirable but the other treatments are also worthwhile, as in some situations the treatments that bring speeds to Safe System levels may not be practical, and an incremental improvement is usually better than none at all.[119]

119 Hillier et al, 2016, p 1.

In short, it is a document full of non-Safe System treatments that work. Most of them may not get speeds down to those required to avoid fatal or serious injuries in a crash, but they are effective in reducing the incidence and severity of crashes. This report doesn't hector us on the need for lower speed limits and ruthless enforcement. Instead, it takes old-fashioned road safety engineering (the basis of road safety audits, crash investigation and safe road design) and uses that experience to offer effective engineering treatments to manage speeds and reduce risks. If rebadging it as 'Safe System' is what it takes to get it published these days, then so be it.

Similarly, a more recent Austroads Research Report AP-R556-17 *Understanding and Improving Safe System Intersection Performance* (Jurewicz et al, 2017) contains much good, sensible stuff like ways to physically contain speeds at critical locations and modify intersection angles.

These two reports — plus AP-R498-15 *Improving the Performance of Safe System Infrastructure: Final Report* (Jurewicz et al, 2015) — illustrate that if we are going to be effective about road safety in a sustained, ongoing way a very big part of that success is going to require spending money on safe infrastructure and making safety improvements to existing infrastructure. Rather than telling road users what to do, we must engineer the roads to elicit safer responses. How refreshing.

When I talk about safe infrastructure and engineering our roads for better safety, it's things like the following, amongst a thousand other treatments:

▶ Relocating hazardous poles.

▶ Installing crash barriers to shield hazards.

▶ Providing or upgrading lighting.

▶ Physically changing the appearance of a road or street so it feels 'walled in'.

▶ Closing median breaks where right turns create problems.

▶ Building roundabouts to reduce speeds, overcome right-angle crashes or assist access.

▶ Banning parking on traffic routes.

▶ Providing refuges, painted medians or pedestrian signals.

▶ Separating oncoming traffic on rural roads.

▶ Limiting access, e.g. through service roads.

▶ Providing separate turning lanes.

In 2017 I made a presentation listing scores of infrastructure-related road safety lessons that have been learnt over the past few decades, only to be largely forgotten.[120]

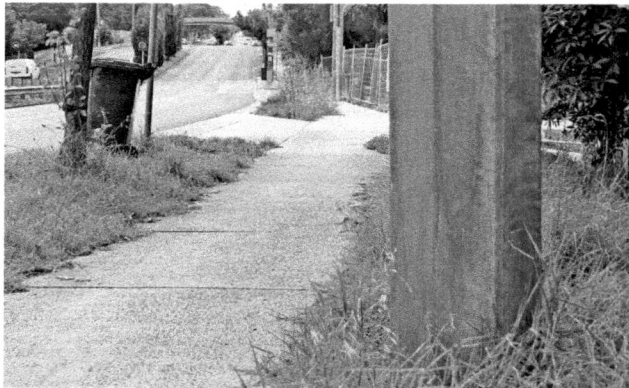

Figure 8.2
Engineering for improved safety: electricity poles have been relocated away from the roadway on the outside of this curve

---

120 Morgan, 2017. Some of the examples appear in this book.

Unfortunately, it's not all enlightenment in this area. As mentioned in Chapter 6, another recent Austroads report focuses on low speed limits as the primary means of reducing risks. This is accompanied by some 'unfortunate' engineering proposals while avoiding the elephant in the room: on-street parking and 'dooring' of cyclists. The report is Austroads Technical Report AP-T330-17 *Safe System Infrastructure on Mixed Use Arterials* (Turner, Partridge et al, 2017) and the example within the report that I wish to discuss is Glen Huntly Road, Elsternwick/Caulfield South (in Melbourne)[121], a four-lane traffic route with trams, through a long strip shopping centre. Parking is time-limited, but there are very few parking bans.

Glen Huntly Road is an arterial road. It used to have a 60 km/h speed limit, though conditions rarely permitted travel speeds that high. A 40 km/h speed limit was implemented a few years ago, from 8 am to midnight, Monday to Saturday, reverting to 60 km/h at other times. The Austroads report includes a table with crash information for this road. A review of the crash data used for that table, plus crash data from before the 40 km/h speed limit (which is not in the Austroads report) is provided in Table 8.1[122].

---

121 As background, VicRoads and its predecessor organisations have never at a corporate level understood functional urban road hierarchies and the need for many important Council roads to function as traffic routes. In recent years VicRoads has surged ahead implementing Network Operating Plans based on a 'SmartRoads' framework, without there being any agreed network of traffic routes upon which to apply this framework. In effect VicRoads has given away large swathes of the required traffic route network across greater Melbourne without ever appreciating there needed to be one. And now they want to play 'Movement and Place' and 'Safe System' on what's left.

122 Table B1 in AP-T330-17 is wrongly titled: it is number of persons injured, not number of crashes. It includes Hawthorn Road intersection. Table 8.1 is number of crashes, between Nepean Highway and Hawthorn Road but excluding both intersections (Hawthorn Road data is excluded due to possible data error in 'after' period: no recorded crashes in final 21 months). 'Before' is from 2 periods: 1.1.01 — 31.12.05 west of Kooyong Road and 1.4.04 — 31.3.09 east from Kooyong Road. 'After' is 9.3.11 — 8.3.16.

Table 8.1

Casualty Crashes, Glen Huntly Road, Elsternwick/Caulfield South

	Motor vehicle only	Motorcycle	Pedestrian	Bicycle	Total in 5 years
Before 40 km/h limit	26	5 (incl. 2 'doored')	15 (15 pedestrians)	16 (incl. 9 'doored')	62
After 40 km/h limit	11	3 (none 'doored')	24 (26 pedestrians)	14 (incl. 5 'doored')	52
Change	Down 60%	Dooring – eliminated Other – no change	Up by 60%	Dooring – down 44% Other – up by 30%	Down 16%

The table shows that the lower 40 km/h speed limit has been very helpful for motorists, helpful for motorcyclists, a mixed blessing for cyclists and a disaster for pedestrians.[123]

In the 'Before' period the single greatest type of collision was cyclists (9) and motorcyclists (2) being hit by the door of a parked car being opened ('dooring'). This has halved but is still a significant problem (with one more crash involving a cyclist hitting a parked car). We know bicycle accidents are under-reported, so these numbers may be just the tip of the iceberg. Yet the 'Indicative [Safe System] Design' in Report AP-T330-17 makes no attempt to deal with this, except to suggest 'Council would need to consider whether [cyclists] are better suited to Glen Eira Road and provide bike lanes there'. That is a parallel arterial road 800 m away that already has bicycle lanes.

---

123 Why the big increase in pedestrian crashes? Possibly it's because not all vehicles travel slower and this adds complexity to pedestrians' decisions. Or perhaps pedestrians think that with the 40 km/h limit it's now 'safe' and caution is no longer needed.

The Indicative Designs/Concept Plans[124] for this road include:

- A 30 km/h speed limit. Perhaps they didn't look at before and after crash numbers for the 40 km/h speed limit?

- Retaining a majority of the on-street parking. Due to the narrow left lane, dooring of cyclists will inevitably continue.

- Suggestions for closing numerous side streets.

- 'Enhanced (and New) Signalised Pedestrian Crossings'. Perhaps they don't know what 'pedestrian crossing' means?

- A raised intersection and approaches with surface treatments at Kooyong Road and at Orrong Road. Yes, they're proposed to be ramped on a tram route.

- A 40 km/h limit on the Kooyong Road approaches, despite nothing in the crash data to warrant it.

- No enhancements (e.g. median islands) to slow turning vehicles where pedestrians have been hit by turning vehicles.

- Removing a pedestrian crosswalk from a signalised intersection and replacing it at new signals 40 m from the intersection. That is likely to increase 'see-through effect' crashes.

- 'Surface treatment' medians to replace painted medians.

Key crash causes were not addressed. This seems to be a theme with the Safe System — that getting into the details through crash investigation is not really that important: 'Crash history is not going to tell us the

---

124 The concept plans in Report AP-T330-17 list the following organisations in their title block: Austroads (as the project funder), Safe System Solutions Pty Ltd, ARRB Group, MWH/Stantec, Monash University Accident Research Centre, Corben Consulting and the University of Adelaide.

risks on the road[125]'. As discussed at the end of Chapter 7, risk equals 'severity of outcome' multiplied by 'likelihood of occurrence', so the crash history in Table 8.1 tells us exactly the risks on this road. Instead, the report offers the population-level 'solution' of an even lower speed limit after the first one was arguably unsuccessful, plus ideas that hint at inexpertise like ramped roadways on a tram route; plenty of coloured surface treatments so they cease being noticed at critical places; and the road's traffic function being given away under the 'Movement and Place' banner. How is this a solution to anything?

Elsewhere in the report, they offer:

► A roundabout with no deflection of vehicle paths (p 39). No deflection = no speed control.

► Advice on one plan that 'chicane helps reduce speeds' whereas it looks like 'chicane helps increase head-on crashes' (p 66), and

► Turning a six-lane 70 km/h divided road into a four-lane divided road with a 50 km/h speed limit (pp 95-101).

This is the direction the collective wisdom of our nation's newfound Safe System specialists is leading us. Please pray for us. Again, it gets back to my 2007 conclusion:

> Where low speed limits are the main thrust of a road safety strategy, it is inevitable that poor design will increase and levels of safety will get worse.

The Safe System is like some government-sanctioned gold rush. The mother lode has been exposed and every self-appointed Safe System expert is on the bandwagon, speeding ahead to get a share. It's the only aspect of the Safe System where there are no proposals to moderate speed. For example, we now have:

---

125 Quoting one of the authors of Turner, Partridge et al, 2017 from a webinar about this report on 8th February 2018.

▶ A road safety audit training course in one city which is now 'Road safety audit: a safe system approach'. The advice I have is that participants of the earlier version of the course were not even pulled up for offering 'solutions' without first identifying what the road safety problem was. So perhaps the name change is quite fitting.

▶ Businesses and products with Safe System in their name.

▶ Road Design departments that are now Safe System Design departments.

If you don't speak Safe System, who's going to listen to you?

My, how perspectives have changed. In the late 1980s Trinca et al stated:

> Progress, if there is to be progress, will be measured by change in personal safety [i.e. crash rates per 100,000 population], and will almost surely produce less safety improvement per dollar.
>
> This is not to say that even the safest country (probably Sweden) has exhausted the possibility for cost-effective improvements, but rather that the measure of such improvement will appear small on the scales used in earlier periods.[126]

In 1985 Sweden's fatality rate per 100,000 population was 9.7. By 1996, prior to adopting Vision Zero and implementing '2+1 with barrier' schemes, it had dropped by 3.4% p.a. to 6.1. By 2015 it was 2.7, a drop (since 1996) of 6.2% p.a.[127]

In the late 1980s, the view was that improvements were going to be slight. These days the view is that zero fatal and serious injury crashes are possible. One view was fairly pessimistic, while the other — in

---

126 Trinca et al, 1988, p 30.

127 ATSB, 2003 and ITF, 2017.

my estimate — is unrealistically optimistic. Accepting a situation or even willing one to happen is no predictor of the future. Both episodes illustrate the need to appreciate that trends are never forever. I expect the future in road safety will be at many places in between these two extreme perspectives.

Let me conclude this chapter on a hopeful note. All the above examples illustrate that if we want to achieve success in reducing crashes (both their numbers and severity) we need to look at the details in the crash data and the details in what are proposed as solutions. And when you get into the details it becomes evident that it's infrastructure and physical treatments that are going to provide long-lasting effective road safety improvements.

The first example is Bell Street in the northern suburbs of Melbourne, described in Case Study 2 in Chapter 2. Although the speed limit was lowered, rather than the specific crash problems being addressed, let's look at a few of those crash problems and what could be done:

- ► The illustrated example of right turns through queued traffic hitting a (non-bus) vehicle in the Bus Lane could be significantly overcome by changing the start of the Bus Lane to a left turn lane, buses excepted, plus some additional signs. The problem could be completely overcome by signal control of the intersection, including removal of the nearby pedestrian signals as part of the project.

- ► Two other locations with right turners colliding with oncoming traffic could be fixed by closing two median breaks and redirecting turning traffic via nearby signalised intersections that already have full right turn control.

- ► One signalised intersection had several crashes where vehicles went through the red light. It turned out to be the only intersection approach without a mast arm (cantilevered) traffic signal. Solution: install the mast arm traffic signal.

▶ At two signalised intersections, right turn crashes on the side road could be overcome by changing the lane allocation and fully controlling the right turns (red, yellow and green arrows).

▶ Crashes into parked vehicles could be avoided by banning parking in the left traffic lanes at all times, with indented parking at one group of shops.

None of those problems was addressed by lowering the speed limit from 70 km/h to 60 km/h.

The second example is a paper about managing older drivers' crash risk by Langford & Oxley, 2006. After references to the Safe System, Langford and Oxley set out the details and extent of the problems, by examining various studies. From this they look at countermeasures, observing the following:

▶ Older driver crashes are predominantly an urban problem. They have difficulties negotiating intersections. Intersection design can reduce crash severity, e.g. by use of roundabouts and full signal control of right turns.

▶ Frailty is a major determinant of crash outcomes, so the greater protection provided by newer vehicles assists in reducing severity and improving survivability.

▶ Older drivers tend to drive at lower speeds, so 'speed policies within the Safe System approach are likely to have only a modest impact upon their own speed choices.'

▶ Only a minority of older drivers has reduced levels of fitness to drive, so there are no rational grounds for implementing mandatory age-based testing of driving fitness for the whole group. The large majority of older drivers are demonstrably as safe as or safer than drivers of other ages.

It is from the examination of these details of 'the problem' that effective countermeasures can be implemented and unhelpful 'solutions' can be avoided.

The third example is a series of in-depth Highway Road Safety Reviews in New South Wales. As explained by Smart el al, 2009:

> Highway Road Safety Reviews are comprehensive, multi-disciplinary reviews of safety issues on key transport routes. These reviews involve extensive investigation and consultation including in-depth analyses of the Highway's crash history, route inspections, workshops, and consultative meetings and reporting. The reviews also place emphasis on further improving the coordination and integration of road improvement projects and ensuring the best safety outcome through an integrated program.

> Highway Road Safety [Reviews] are innovative holistic approaches aimed at reducing the road toll by providing low cost, effective safety engineering works and behavioural/technology programs.

> Highway Road Safety Reviews have been carried out for the Pacific (in 2003/04) and Princes Highways (in 2004/05). These resulted in the implementation of a 3 year, $35M road safety program for the Pacific Highway and a 3 year, $30m program of safety improvements for the Princes Highway. The strategy resulted in substantial reductions in both fatal and injury crashes with a first-year post-implementation saving of 50 lives and 164 injuries on both routes.

> Because of the success of previous highway safety reviews, it was decided to undertake a similar safety review of the Newell Highway.

On the Newell Highway the in-depth, multi-disciplinary investigation resulted in the following key findings:

▶ The largest proportion of fatal crashes is off path crashes or rollover crashes.

106

- ▶ Fatigue and high speeds were considered to be major behavioural factors in casualty crashes.

- ▶ Almost 30% of all fatal crashes were head-on crashes (13 out of 38), with heavy vehicles being involved in 92% of these head-on crashes.

- ▶ Heavy vehicle involvement in four of the five head-on fatal crashes in 2006.

- ▶ Heavy vehicle and interstate [contractors] accounting for a high proportion of fatal crash involvements, however, recorded alcohol involvement in crashes was underrepresented compared to the incidence on other country highways.

- ▶ Earlier studies identified curves within the 200 m to 600 m radius range are over-represented in fatal and serious injury crashes. There are 104 such curves on the Newell Highway.

It was from the detailed investigation of the road's layout and features and the crashes that the following actions were proposed:

Enhanced Road Safety Engineering Works

- ▶ Junction treatments: sealed side road approaches, upgraded warning signs, better delineation and separation of movements at major intersections.

- ▶ Road environment: clear zones of at least 6 metres or installation of wire rope safety barriers.

- ▶ Widened tactile centreline: at two locations to separate opposing traffic by approximately 1.0 m, still allowing vehicles to legally overtake at safe locations.

- ▶ Road alignment: shoulder widening and sealing, especially around the outsides of curves, delineation around the outsides of curves, realign or re-camber some curves.

Behavioural Programs

- ▶ Driver fatigue: targeted public education campaigns, use of VMSs, approach advice about truck stops.

- ▶ Speeding drivers: public education campaigns, enforcement at key locations, use of VMSs.

- ▶ Pedestrian safety: 60 km/h limits in towns reduced to 50 km/h (though no information was offered about actual crashes in the 60 km/h zones).

Enforcement Programs

- ▶ Police enforcement.

- ▶ Heavy vehicle enforcement: random mobile enforcement.

The value of in-depth investigation and analysis is obvious from the crash reduction results on the two highways reviewed before the Newell Highway. The Newell Highway example has illustrated how much of the safety improvements are engineering-based and infrastructure improvements.

In view of these three examples, it is odd to note that in the 160-odd pages of their book on future directions for road safety, Johnston et al spend only half a page talking about road design and infrastructure.[128]

Let me now undertake a review of my own, drawing together the experience discussed so far.

---

128 Johnston et al, 2017, pp 137-138.

Chapter 9

AUSTRALIA'S SAFE SYSTEM — A REVIEW

The Safe System has as its aim the elimination of fatal and serious injuries on our roads. This element came from Sweden's Vision Zero, which I discussed in Chapter 3. As Australia's national road safety strategy[129] puts it:

No person should be killed or seriously injured on Australia's roads

Unfortunately, as explained in Chapter 5, a focus on fatal and serious injury crashes is likely to be counterproductive; a strategy that seeks to address all crashes is likely to be more successful in reducing the total number of fatal and serious crashes than the Safe System which addresses just one percent of all crashes. Worse still, in Chapter 2 it was concluded from experience that the pursuit of 'zero' will lead to a return to the bad old days of blaming the driver, with fewer safe infrastructure treatments being implemented and the actual causes of crashes not being addressed.

Under the Safe System the elimination of fatal and serious crashes is expected to ultimately come about through the management of kinetic energy: any collisions of vehicles with other vehicles, other road users or objects should not result in the transfer or dissipation of energy at a rate that will result in serious or fatal injuries to a human body. And so at the heart of the Safe System is the 'Human tolerance to physical force' (see Figure 9.1).

---

129 Australian Transport Council, 2011, p 31.

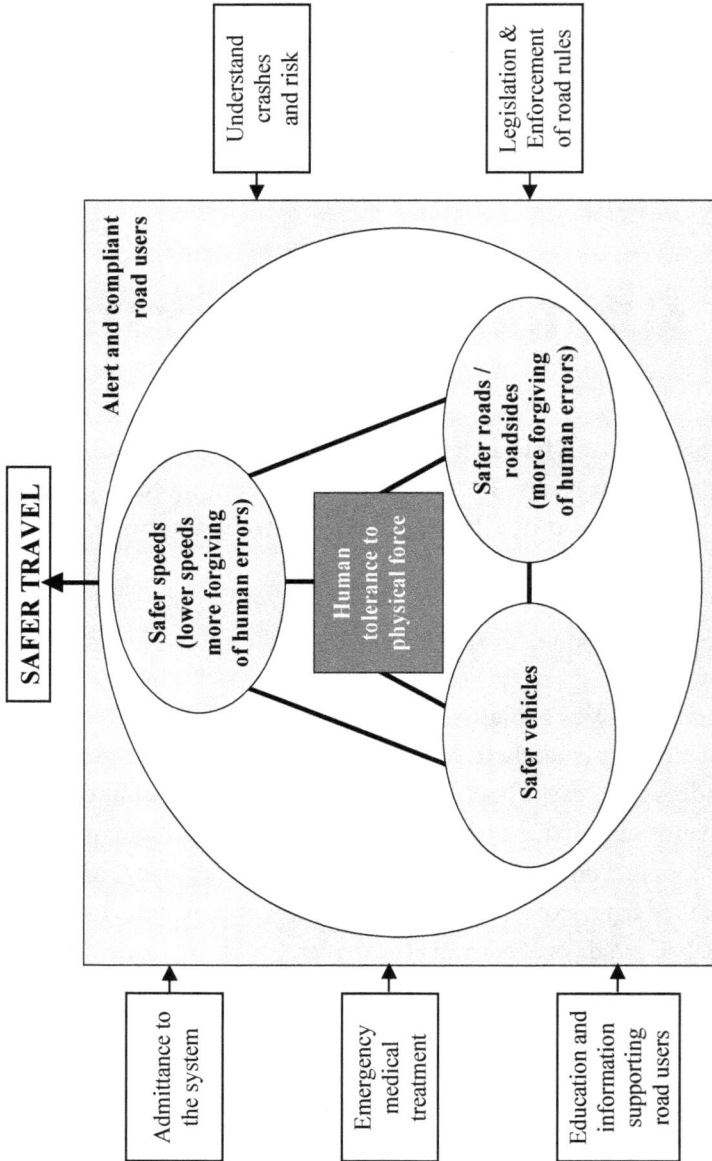

Figure 9.1
Australia's Safe System[130]

---

130 Same image as Figure 1.1. Based on Figure 6.5 in Johnston et al, 2017.
    Attributed to Eric Howard. Similar diagram appears in OECD/ITF, 2008
    at p 114 as 'Australia's Safe System — Illustrating the primacy of speed
    management' without *Emergency medical treatment.*

Seat belts and flexible roadside barriers are designed to reduce the rate of energy exchange for this purpose. But they are existing technologies. As speed is a major component of kinetic energy, a major plank of the Safe System is speed management. But this is invariably thought of in terms of speed limits and enforcement, rather than using engineering to achieve desired speeds, as is done in the Netherlands.

This *speed limits* approach is not very hopeful, given the discussion at the end of Chapter 5 and given that more than once in this book we have seen that where low speed limits are the main thrust of a road safety initiative, it is accompanied by poor design and levels of safety that are worse than they could be. This approach to speed management inevitably reduces to pointlessly low speed limits and automated enforcement.

Through this book I have made several points about speed and safety and the concerns I have with the Safe System's approach to speed. At the end of this chapter I attempt to draw together and summarise the main points, so they are all in the one place.

According to our national strategy, the Safe System approach:

> accepts that people using the road network will make mistakes and therefore the whole system needs to be more forgiving of those errors.[131]

This infers that we should let them keep making those mistakes and just arrange things differently so the consequences are less severe (no fatal or serious injuries). This is dealing with only half the issue. Stopping people making mistakes in the first place must be part of any effective road safety strategy. Yet it is absent from the Safe System.

Looking at Figure 9.1 the three 'pillars' of the Safe System are safer roads and roadsides, safer vehicles and safer speeds. Virtually the whole area of road user behaviour has been ignored: the only aspect of behaviour

---

131 Australian Transport Council, 2011, p ii.

of any interest in the original Victorian 2004 Safe System framework is speeds. This more than anything else illustrates the obsession with speed in the origins of the Safe System. It is interesting to see that since 2004 terms like *Safe Road Use* and *Safer People* have been introduced as an additional pillar of the Safe System. For example, see the New Zealand Safe System framework in Figure 9.2. It looks like sense has prevailed and road user behaviour has been reintroduced. Alas, this is not so; these new terms are simply the reframing of 'alert and compliant road users' into an additional pillar.[132] Road user behaviour is still absent.

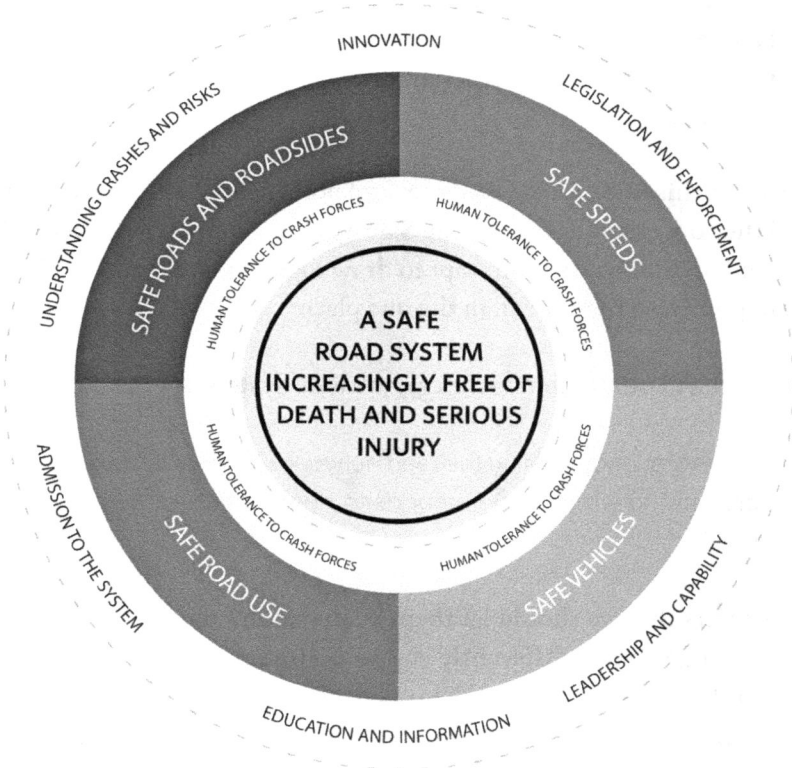

Figure 9.2
The New Zealand Safe System framework
(Image courtesy of NZ Transport Agency and the Ministry of Transport)

---

132 Turner & Jurewicz, 2016, p 1.

An understanding of road user behaviour is essential for achieving effective road safety improvements. Ogden, 1996 provides an extensive overview of the topic in his chapter *Human factors in road traffic*, which includes discussion about the vital topic of human decision making[133] and references to the work of others. Virtually all of this is ignored by the Safe System. This is of course understandable: the Safe System needs road users to be 'alert and compliant', rather than what they really are. It proposes lower and lower speed limits and ignores the scientific evidence about how road users will respond.

'Safer vehicles' covers a wide area of activity. For example, see Grzebieta & Rechnitzer, 2001. It is outside my area of expertise but it is important to acknowledge the great reductions in crash trauma that have been achieved over the years through features like anti-lock braking, collapsible steering columns, head restraints and seat belts, plus many others. Autonomous braking offers future hope.

'Safer roads' is not just about the Safe System's simplistic message of 'roads and roadsides more forgiving of human error'. Quayle, 2015 notes that during the time of the Federal Office of Road Safety (FORS) they reduced safety in road design to four basic principles that can be put into effect on all classes of road:[134]

- ▶ Minimising the potential for collision.

- ▶ Ensuring relatively consistent design standards so that drivers are not presented with unexpected situations.

- ▶ Providing accurate information to road users.

- ▶ Providing a forgiving roadside.

So far as I can see the Safe System only deals with the last of these

---

133 If we understand human decision making we can help road users avoid making mistakes.

134 Quayle, 2015, p 170.

four. FORS worked out long ago something the proponents of the Safe System fail to realise: a safe road system is about much more than just kinetic energy.

I've previously described the Australian Safe System's demand for 'alert and compliant road users' as road safety fundamentalism: 'Acceptable drivers all must be — mild, obedient, good as He' (Morgan, 2014). Complying with road rules and other laws is all very well, so long as those laws are appropriate and the penalties proportionate. The Safe System input of 'Enforcement of road rules' suggests that the law is some benign entity that has been created with everyone's best interests in mind. It isn't; many road infringement penalties are completely out of proportion to the risk or danger involved. Any framework for effective road safety needs to acknowledge this and require that safety-related laws be based on evidence of need and effectiveness.

The New Zealand Safe System avoids promising the ultimate elimination of fatal and serious injuries. Instead, it proposes a very low death toll, with serious injuries being increasingly rare (see Figure 9.2).

And what about resources? The Safe System is silent on this and what we are seeing is the Safe System being used as an excuse for not applying adequate resources to road safety, both in funding of effective infrastructure improvements and in skill resources. According to the Safe System, if funds aren't available to improve infrastructure, the speed limit needs to be lowered. It's an appealing circular argument for any government: safer infrastructure costs money, but lowering speed limits is cheap. Then we can increase enforcement. That lets us show we're serious about road safety and we get revenue from the speeding fines to offset government costs for the option that was cheaper than building safer infrastructure!

And that's what happens: the Safe System leads to fewer safety improvements.

In 1988, before the Safe System, Trinca et al stated:

Traffic crashes and their consequences can be reduced — even in countries with the lowest traffic injury rates at present — by the systematic and widespread application of current knowledge in the fields of vehicle design, road design and traffic management supported by appropriate regulatory controls. Countermeasure application often lags behind our knowledge of what does and does not work and it is instructive to examine why.[135]

They then went on to discuss the contemporary reasons why countermeasures were not being applied. We could now add to their list the loss of one of the most important resources of all — experience, expertise and skills in state road authorities. Consider Case Study 3.

CASE STUDY 3
Road Authority Skills

Back when skills were still plentiful, a state road authority regional office identified a crash problem at a high-speed rural intersection. A reasonably busy side road intersected with a major divided road which had a very wide median. Approaching on the straight main road, the side road was on the left, just after the main road began curving right. The main road had a separate left turn lane. The crash problem involved drivers exiting the side road (to reach the other carriageway of the main road) colliding with a vehicle travelling through on the main road. The speed limit had been dropped from 100 km/h to 90 km/h without great effect. Probably because their skills were plentiful, the road authority engineers sought to add more skills to the assessment and involved an independent road safety engineer. From the resultant discussions, site investigation and assessment it was determined that vehicles in the left turn deceleration lane were hiding following vehicles in the left-hand through lane, leading drivers in the side road to think the through lanes were clear. It was evident that older drivers were dealing less well with this because the 'hiding' of

---

135 Trinca et al, 1988, p 44.

through vehicles added a level of complexity to their decision-making. From this investigation the conventional left turn lane was changed to a tapered exit, so left turners were outside the sight line between side road drivers and the through traffic lanes (see Figure 9.3). The crash problem disappeared overnight.[136]

It would appear that when a road authority has expertise, it knows when to seek the assistance of other experts. With the loss of expertise, people in road authorities are less likely to seek expert advice simply because they don't recognise situations where that advice may help (They 'don't know what they don't know'). Consequently, fewer crash problems are fixed. Any effective 'system' for reducing the incidence and severity of crashes must include sufficient numbers of skilled and experienced engineering practitioners directly employed in our state and territory road authorities. These days we do not have them and that is one reason why the Safe System has taken hold.

Figure 9.3
The new straight, tapered left turn exit, clear of the sight line from the side road. The old left turn lane was parallel to the through lanes.

---

136 It is worth noting that a similar problem on a major highway in another state was not solved by this type of left turn lane realignment because the main road approach was on a relatively tight curve and vehicles in the left-hand through lane continued to hide vehicles in the right-hand through lane.

There are many other shortcomings with the Australian Safe System. These are set out in Figure 9.4, together with the issues mentioned so far. The Safe System framework illustrated in Figures 1.1 and 9.1 is heavily distorted towards telling people what to do and exerting external control on the basis that the experts know best. There is no trusting people or encouraging and supporting them. It concerns itself with little about road user behaviour and ignores the experience behind Figure 2.1. It pays only lip service to engineering, infrastructure and physical solutions. The fact that many good research reports have recently been produced covering 'Safe System infrastructure' probably has more to do with level heads seeing an opportunity to promote safe infrastructure options than the proponents of the Safe System having any appreciation of this aspect.[137]

---

137 As mentioned in Chapter 8, Johnston et al, 2017 spend only half a page talking about road design and infrastructure in their 160-page book.

Notes for Figure 9.4

1. The Safe System framework at the centre of this diagram is an Australian construct. Redrawn, based on Fig. 6.5 in Johnston et al, 2017. Attributed to Eric Howard, 2004.

2. Other variations exist, including some quite different ones in other countries.

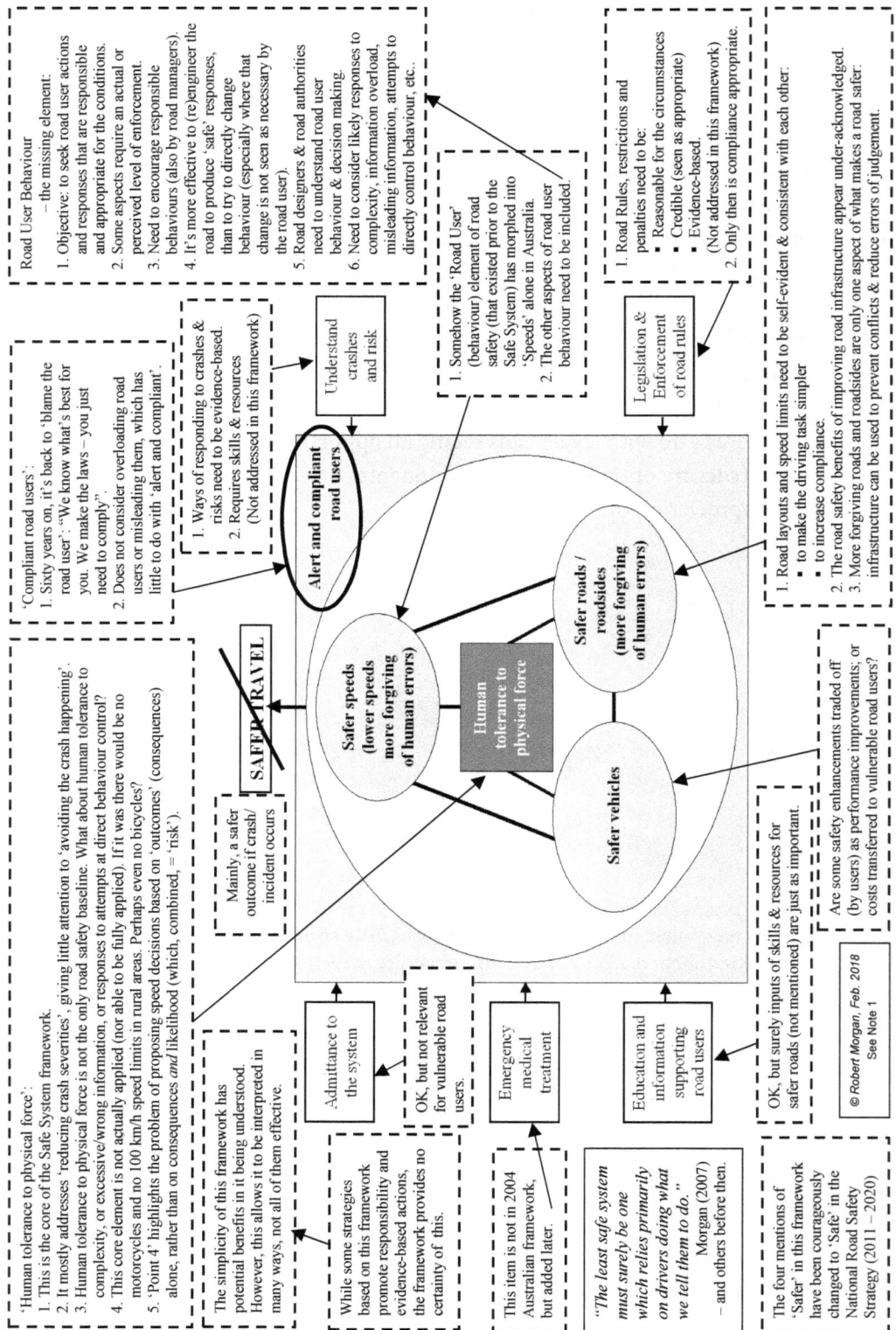

Road User Behaviour
– the missing element:
1. Objective: to seek road user actions and responses that are responsible and appropriate for the conditions.
2. Some aspects require an actual or perceived level of enforcement.
3. Need to encourage responsible behaviours (also by road managers).
4. It's more effective to (re)engineer the road to produce 'safe' responses, than to try to directly change behaviour (especially where that change is not seen as necessary by the road user).
5. Road designers & road authorities need to understand road user behaviour & decision making.
6. Need to consider likely responses to complexity, information overload, misleading information, attempts to directly control behaviour, etc.

1. Somehow the 'Road User' (behaviour) element of road safety (that existed prior to the Safe System) has morphed into 'Speeds' alone in Australia.
2. The other aspects of road user behaviour need to be included.

Legislation & Enforcement of road rules

1. Road Rules, restrictions and penalties need to be:
   ▪ Reasonable for the circumstances
   ▪ Credible (seen as appropriate)
   ▪ Evidence-based
   (Not addressed in this framework)
2. Only then is compliance appropriate.

1. Road layouts and speed limits need to be self-evident & consistent with each other:
   • to make the driving task simpler
   • to increase compliance.
2. The road safety benefits of improving road infrastructure appear under-acknowledged.
3. More forgiving roads and roadsides are only one aspect of what makes a road safer: infrastructure can be used to prevent conflicts & reduce errors of judgement.

'Compliant road users':
1. Sixty years on, it's back to 'blame the road user': "We know what's best for you. We make the laws — you just need to comply".
2. Does not consider overloading road users or misleading them, which has little to do with 'alert and compliant'.

1. Ways of responding to crashes & risks need to be evidence-based.
2. Requires skills & resources (Not addressed in this framework)

Understand crashes and risk

Alert and compliant road users

'Human tolerance to physical force':
1. This is the core of the Safe System framework.
2. It mostly addresses 'reducing crash severities', giving little attention to 'avoiding the crash happening'.
3. Human tolerance to physical force is not the only road safety baseline. What about human tolerance to complexity, or excessive/wrong information, or responses to attempts at direct behaviour control?
4. This core element is not actually applied (nor able to be fully applied). If it was there would be no motorcycles and no 100 km/h speed limits in rural areas. Perhaps even no bicycles?
5. 'Point 4' highlights the problem of proposing speed decisions based on 'outcomes' (consequences) alone, rather than on consequences *and* likelihood (which, combined, = 'risk').

Safer speeds (lower speeds more forgiving of human errors)

Safer roads / roadsides (more forgiving of human errors)

Human tolerance to physical force

SAFER TRAVEL

Safer vehicles

Mainly, a safer outcome if crash/ incident occurs

Are some safety enhancements traded off (by users) as performance improvements; or costs transferred to vulnerable road users?

The simplicity of this framework has potential benefits in it being understood. However, this allows it to be interpreted in many ways, not all of them effective.

While some strategies based on this framework promote responsibility and evidence-based actions, the framework provides no certainty of this.

Admittance to the system

OK, but not relevant for vulnerable road users.

Emergency medical treatment

Education and information supporting road users

OK, but surely inputs of skills & resources for safer roads (not mentioned) are just as important.

This item is not in 2004 Australian framework, but added later.

"The least safe system must surely be one which relies primarily on drivers doing what we tell them to do." Morgan (2007) – and others before then.

The four mentions of 'Safer' in this framework have been courageously changed to 'Safe' in the National Road Safety Strategy (2011 – 2020)

© *Robert Morgan, Feb. 2018* See Note 1

Figure 9.4
Critique of the Safe System Framework

Speed and safety — a summary

Let me draw together the points I've made through this book about speed and safety and also make some conclusions. Speed management is an important part of creating a safer road network — so we need to be effective in what we do.

85th percentile speeds

► The 85th percentile speed is the speed at or below which 85% of drivers travel, measured under free flow conditions (Chapter 7).

► A speed limit set at the 85th percentile speed results in the lowest amount of speed variation. It has the greatest credibility. It requires the lowest levels of enforcement (Chapter 7).

► As drivers typically select a travel speed based on the road and roadside environments and activity on the road and on adjacent roadside areas, these are the factors that typically determine 85th percentile speeds.

► The speed limits in Table 7.1 generally approximate 85th percentile speeds in typical situations on Australian roads.

► Simply because signs advise there is a particular speed limit, that is no guarantee travel speeds will match that limit (start of Chapter 7). For example, there are numerous examples of speed limits being raised and mean travel speeds not changing (Chapter 7).

► Speed limits based on 'the level of protection' (i.e. the Safe System approach), rather than on the road, roadside and traffic conditions are likely to be significantly lower than the 85th percentile speed, except on a high-standard freeway (Chapter 6).

► Speed limits lower than the 85th percentile speed typically result in a wider variation in travel speeds along the road (Chapter 2).

► Speed limits lower than the 85th percentile speed typically require a greater level of enforcement and result in a greater number of drivers receiving speeding penalties.

► Lower travel speeds are associated with lower crash risk. Countering this, wider speed variation is associated with greater crash risk. Thus, where an unreasonably low speed limit (i.e. one for which the purpose is not apparent, so drivers who 'drive to the conditions' adopt a higher speed) is implemented, the speed variation effect will reduce the lower travel speed benefit. In the example of changes between 60 and 70 km/h speed limits on divided urban roads with access driveways, one effect appears to cancel the other effect (Chapter 2).

Risk of crashes with higher speeds

► Travelling at a significantly higher speed than other vehicles is associated with a significantly higher risk of being involved in a casualty crash. This effect increases exponentially (second half of Chapter 5).

► Travelling at a slightly higher speed than other vehicles is associated with virtually no increase in the risk of being involved in a casualty crash[138]. Kloeden et al, 1997 assert that 'In a 60 km/h speed limit area, the risk of involvement in a casualty crash doubles with each 5 km/h increase in travelling speed above 60 km/h.' In stark contrast, Bobevski et al, 2007 — studying the consequences of Victoria's *Wipe off 5* campaign and the associated reduction in speed enforcement tolerance — established that with a 5 km/h reduction in mean and 85th percentile travel speeds there was no significant change in

---

138 The risk could be said to change from extremely low to almost extremely low.

casualty crashes. I conclude that with the Kloeden et al study, although there was a correlation between speed and casualty crash involvement, that's all it was: a correlation, not cause and effect. The speed was not the cause of the casualty crash involvement (first half of Chapter 7).

▶ Enforcement policies that focus on population-level shifts in travel speed (see below) should be viewed in this context.

Speeds and crashes

▶ While lower travel speeds are associated with lower crash risk, lowering the speed limit will not automatically lower travel speeds (start of Chapter 7).

▶ Lowering the speed limit will not automatically reduce crashes. The actual causes of particular crashes need to be identified and countermeasures implemented that target the actual crash causes (Chapters 2 and 8).

Seeking to directly change driver behaviour

▶ Reducing a speed limit below the 85th percentile speed is an example of seeking to directly change driver behaviour. The behaviour change that is sought is lower travel speed.

▶ (Re-engineering or changing the road environment to slow drivers down results in a new, lower 85th percentile speed, not a speed that is lower than the 85th percentile speed).

▶ Attempts at directly changing driver behaviour are typically ineffective unless there is a perception of significant enforcement (Chapter 2). Such perception is based on factors which include observed levels of actual enforcement.

▶ Using a lower speed limit in order to directly change driver behaviour (i.e. to achieve lower travel speeds) is an example of seeking a population-level shift in behaviour (second half of Chapter 5).

▶ (From public health) reducing risk amongst a population can be attempted by either seeking a population-level shift in behaviour amongst the majority who have a low risk, or seeking behaviour change by 'the few' who have a high risk (second half of Chapter 5).

▶ With drinking and driving, because crash risk increases exponentially with increases in consumption, it has been established that it is more effective to target 'the few' who are over a reasonable limit (BAC of 0.05) than to seek a population-level change in behaviour by reducing the 'reasonable limit' (second half of Chapter 5).

▶ Because casualty crash risk from speed is a similar phenomenon (risk increases exponentially with increases in travel speed) it follows that it is more effective to target 'the few' who are over a reasonable limit (i.e. the 85th percentile speed plus a reasonable enforcement tolerance) than to seek a population-wide change of behaviour by reducing the limit and/or reducing the enforcement tolerance below reasonable levels (second half of Chapter 5 and first half of Chapter 7).

▶ Where a population-wide travel speed reduction is needed because there are problems of speed-related risk (speed-related crashes), it will be more effective to re-engineer the road to achieve a lower 85th percentile speed, than to rely on a lower speed limit.[139]

---

139 This is akin to the Netherlands' Sustainable Safety program and their 'self-explaining road' concept (see the end of Chapter 3). An additional benefit of this approach is that if low speed limits are used very sparingly, those low limits that *are* used are likely to be more effective because the overall system will have greater credibility.

▶ These points favour the use of 85th percentile speeds as the significant primary factor to be used in determining speed limits for safety, not just for mobility.

The Safe System

▶ The Safe System is completely at odds with these points.

Chapter 10

ROAD SAFETY NEEDS A STAR SYSTEM, NOT A STALINIST SYSTEM

Effective road safety is more than just kinetic energy, low speed limits and automated enforcement.

As well as the road, the vehicle and speed, it needs to consider all aspects of road user behaviour and the effects of laws and policies. Accident blackspot investigation and treatment may only deal with a small part of our road network but the experience practitioners gain from it is vital for the safe design and management of the whole road network.

Instead of the Safe System with its moralising deceit that 'zero is possible', its generally miserable attitude towards road users and its focus on behaviour control that Stalin would have been proud of, we need an alternative that values and respects people, seeks to reduce mistakes as well as reducing the consequences of mistakes and focuses on achieving ongoing long-term effective road safety improvements.

That alternative is shown in Figure 10.1. I have chosen to call it the SAFETY STAR SYSTEM.

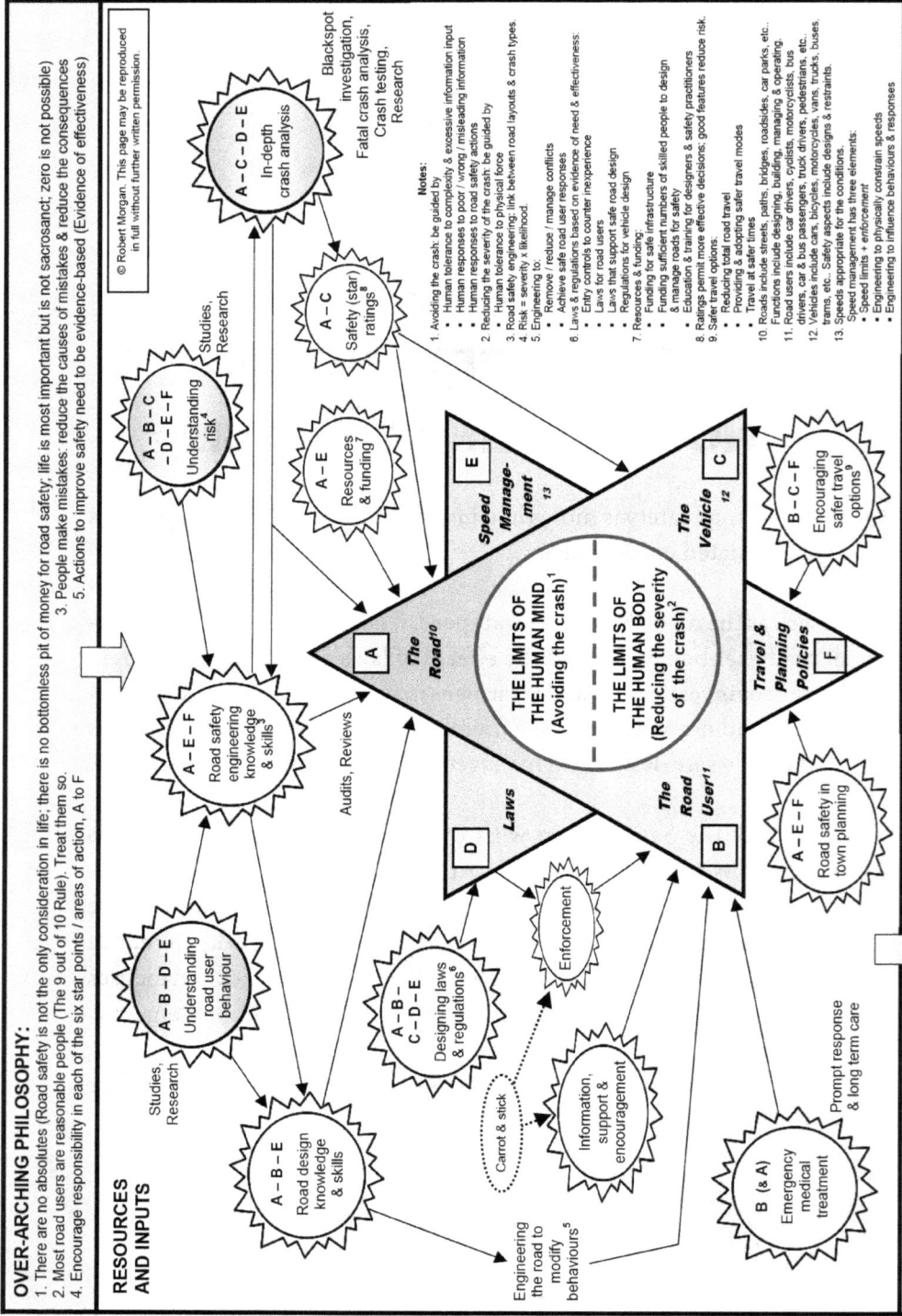

OVER-ARCHING PHILOSOPHY:
1. There are no absolutes (Road safety is not the only consideration in life; there is no bottomless pit of money for road safety; life is most important but is not sacrosanct; zero is not possible)
2. Most road users are reasonable people (The 9 out of 10 Rule). Treat them so.
3. People make mistakes. Reduce the causes of mistakes & reduce the consequences
4. Encourage responsibility in each of the six star points / areas of action, A to F
5. Actions to improve safety need to be evidence-based (Evidence of effectiveness)

RESOURCES AND INPUTS

Blackspot investigation, Fatal crash analysis, Crash testing, Research

A – C – D – E In-depth crash analysis

A – B – C – D – E – F Understanding risk[4]

Studies, Research

A – C Safety (star) ratings[8]

A – E Resources & funding[7]

A – E – F Road safety engineering knowledge & skills[3]

Audits, Reviews

Studies, Research

A – B – D – E Understanding road user behaviour

A – B – E Road design knowledge & skills

A – B – C – D – E Designing laws & regulations[6]

Engineering the road to modify behaviours[5]

Enforcement

Carrot & stick

Information, support & encouragement

B (& A) Emergency medical treatment

Prompt response & long term care

B – C – F Encouraging safer travel options[9]

A – E – F Road safety in town planning

Star centre triangle:

E Speed Management[13]

A The Road[10]

C The Vehicle[12]

THE LIMITS OF THE HUMAN MIND (Avoiding the crash)[1]

THE LIMITS OF THE HUMAN BODY (Reducing the severity of the crash)[2]

D Laws

B The Road User[11]

F Travel & Planning Policies

Notes:
1. Avoiding the crash be guided by
   - Human tolerance to complexity & excessive information input
   - Human responses to poor / wrong / misleading information
   - Human responses to road safety actions
2. Reducing the severity of the crash be guided by
   - Human tolerance to physical force
3. Road safety engineering: link between road layouts & crash types.
4. Risk = severity x likelihood
5. Engineering to:
   - Remove / reduce / manage conflicts
   - Achieve safe road user responses
6. Laws & regulations based on evidence of need & effectiveness:
   - Entry controls to counter inexperience
   - Laws for road users
   - Regulations for vehicle design
7. Resources & funding:
   - Funding for safe infrastructure
   - Funding sufficient numbers of skilled people to design & manage roads for safety
   - Education & training for designers & safety practitioners
8. Ratings permit more effective decisions: good features reduce risk.
9. Safer travel options:
   - Reducing total road travel
   - Providing & adopting safer travel modes
   - Travel at safer times
10. Roads include streets, paths, bridges, roadsides, car parks, etc. Functions include designing, building, managing & operating.
11. Road users include car drivers, cyclists, motorcyclists, bus drivers, car & bus passengers, truck drivers, pedestrians, etc.
12. Vehicles include cars, bicycles, motorcycles, vans, trucks, buses, trams, etc. Safety aspects include designs & restraints.
13. Speed management has three elements:
    - Speed limits + enforcement
    - Engineering to physically constrain speeds
    - Engineering to influence behaviours & responses

© Robert Morgan, January 2018

OUTCOMES: Fewer crashes; Less severe results from crashes

'The Safety Star System (The Morgan Star)'

Figure 10.1
The Safety Star System

Notes for Figure 10.1

1. Avoiding the crash: be guided by
   - ► Human tolerance to complexity & excessive information input
   - ► Human responses to poor / wrong / misleading information
   - ► Human responses to road safety actions
2. Reducing the severity of the crash: be guided by
   - ► Human tolerance to physical force
3. Road safety engineering: link between road layouts & crash types.
4. Risk = severity x likelihood.
5. Engineering to:
   - ► Remove / reduce / manage conflicts
   - ► Achieve safe road user responses
6. Laws & regulations based on evidence of need & effectiveness:
   - ► Entry controls to counter inexperience
   - ► Laws for road users
   - ► Laws that support safe road design
   - ► Regulations for vehicle design
7. Resources & funding:
   - ► Funding for safe infrastructure
   - ► Funding sufficient numbers of skilled people to design & manage roads for safety
   - ► Education & training for designers & safety practitioners
8. Ratings permit more effective decisions; good features reduce risk.
9. Safer travel options:
   - ► Reducing total road travel
   - ► Providing & adopting safer travel modes
   - ► Travel at safer times
10. Roads include streets, paths, bridges, roadsides, car parks, etc.. Functions include designing, building, managing & operating.
11. Road users include car drivers, cyclists, motorcyclists, bus drivers, car & bus passengers, truck drivers, pedestrians, etc ..
12. Vehicles include cars, bicycles, motorcycles, vans, trucks, buses, trams, etc.. Safety aspects include designs & restraints.
13. Speeds appropriate for the conditions. Speed management has three elements:
    - ► Speed limits + *enforcement*
    - ► Engineering to physically constrain speeds
    - ► Engineering to influence behaviours & responses

The Philosophy

The over-arching philosophy of the Safety Star System has five elements:

1.   There are no absolutes

Life is complex. Simplistic slogans like 'zero is possible' are a deceit. Road safety is not the only consideration in life. Of course life is important and we all seek to preserve it but suggesting it is 'sacrosanct' will inevitably lead to perverse decisions that impinge on other values we hold dear. In a fair world no one would die or suffer serious injury on our roads. But the world is not always fair and we should seek to do the best we can in an organised, systematic and effective way, based on evidence. Likewise, there is no bottomless pit of money for road safety. Greater expenditure is needed, but it needs to be effectively applied, especially on long-lasting improvements like safer infrastructure and engineering improvements.

2.   Most people are reasonable

All people deserve respect. Most people are reasonable (the 9 out of 10 rule) and deserve to be trusted. Others appear on TV shows like *Highway Patrol* and *Border Stupidity*. Traffic management systems need to account for them, but that should not be the main focus of a road safety strategy — treating us all like we are miscreants until proven innocent — just because authorities don't know how to deal with the problem 10%.[140] With the likelihood of an all-pervasive network of automated speed cameras enforcing unreasonably low speed limits under the Safe System, everyone will be treated like a criminal who needs constant surveillance. This is just lazy policy based on the ultimate in criminal profiling: we are all viewed as potential criminals. Any effective road safety strategy needs to support the majority and encourage them to make appropriate and effective decisions as they negotiate the road system.

---

140 As mentioned near the end of Chapter 5, one former senior traffic police officer suggested the number is more like 4%.

3.   People make mistakes

People make mistakes. An effective road safety strategy should not focus only on reducing the consequences of those mistakes. It should also seek to help road users avoid making mistakes and not put them in positions where we know the chances of a mistake are high and the consequences are significant. This requires an understanding of human decision making, which is part of understanding road user behaviour.

4.   Responsibility

If people act responsibly in the areas of their life that affect road safety, there will be fewer and less severe crashes. For example, if people drive to the conditions and don't let alcohol affect their driving judgement. But acting responsibly doesn't only apply to road users. For example, law makers need to acquaint themselves with the implications of proposed laws; professionals need to act only within their areas of expertise and their levels of experience. Most importantly, road designers need to make sure there are no surprises for road users, and they need to get road safety engineering expertise input into their designs. Currently the severe loss of skills and experience within state road authorities is preventing this, so taking responsibility needs to start at the top: state governments and state road authority chief executives need to acknowledge the crisis stage this has reached and take action to restore skills and experience in all engineering areas that influence road safety.

5.   Actions to improve road safety need to be evidence-based

With any action to improve road safety, there needs to be evidence of its effectiveness. This book has illustrated examples where evidence is weak or inconclusive or conflicting. These limitations need to be acknowledged. And where the evidence is simply not there or experience shows that particular benefits will not happen, those actions should not proceed. Actions that are known to be effective should be introduced when opportunities arise, if they are cost-effective: a road safety strategy can't be based on 'if we can only save one life here it will be worthwhile'. Extending one conclusion beyond its applicability is not applying evidence: for example, if a two way (undivided) rural road experiences a drop in crashes when the speed limit is reduced by 10 km/h from 110 km/h, that does not mean a 10 km/h limit drop on a 70 km/h urban divided road will see the same benefits. And evidence needs to be more than just evidence of crash reduction: if the evidence is that a safety program has negative outcomes in other ways, that needs to be considered as well.

The Core Issues

While the Safe System has at its core 'the limits of the human body to withstand physical forces' (the kinetic energy aspects), this is only half the issue. The other half is the limits of the human mind to deal with information and other inputs. This includes complex, excessive, poor, wrong or misleading information. If the information being received by road users as they travel is factually correct, appropriate and within the limits of the brain to process it (see Figure 3.1), then mistakes and errors of judgement can be avoided and the risk of collisions is reduced.

The Safety Star System has at its core both:

► The limits of the human mind, so crashes can be avoided, and

► The limits of the human body, so the severity of the crash can be reduced.

CASE STUDY 4

'The Dangerous Intersection'
— exceeding the limits of the human mind to cope

Early one night a truck driver was travelling through a semi-rural area. He had not been on this road before, there was no other traffic around and he had been told that at the end of the road he needed to turn right. He had not been drinking. He was travelling a little slower than the 80 km/h speed limit. He was approaching a T-intersection on his left. He was not aware of this and he would later say that he didn't see the side road warning sign. The street light at the corner was out, but other street lights further along the road were illuminated, in a straight line. At the same time a car driver was approaching the intersection along the side road. To the truck driver's horror this car driver came out into the intersection without stopping, turning right. The truck driver swerved to avoid a crash, but instinctively swerved to the right and hit the car, resulting in fatal injuries to two of the car occupants.

What the truck driver did not know was that the Council had changed the priority at the corner. The road straight ahead had been closed off; all the *through traffic* turned at this corner and so Council put the intersection priority that way, nine years earlier. This had not been entirely successful: there had been countless crashes and local people referred to it as 'the dangerous intersection'.

Figure 10.2
View approaching in the direction of the truck driver
(Image courtesy of Jamieson Foley Pty Ltd. Modified by the author to replicate
conditions at the time of the accident)

Figure 10.2 shows the approach by the truck driver, including the side road warning sign. On the sign the side road and the bottom of the T are thicker than the top of the T. Had the truck driver seen the sign, what might it have told him? Simply that there's a side road? The important road is the side road? Most traffic goes around the corner? The priority goes around the corner? Perhaps. But what the sign does not even hint at is that if you proceed straight ahead you must give way to any traffic from the side road. Yet this was the most critical road safety message an approaching driver needed to understand.

Figure 10.3
View closer to the intersection
(Image courtesy of Jamieson Foley Pty Ltd. Modified by the author to replicate
conditions at the time of the accident)

What else might have suggested a need to give way? Figure 10.3 is closer to the intersection. The No Through Road sign and the back of the Give Way sign offer no hint. The centre line goes around the corner, but it has a long gap[141], so there is only a short section that curves across the driver's path. In any event, by the time that curved line is observed it's too late.

So there was nothing to alert unfamiliar drivers that conditions were different and that they would need to give way if travelling straight ahead. The information they received was misleading. Similarly, drivers from the side road were being misled into thinking it was safe to take the priority the Council gave them. As a consequence, crashes had been happening for years.

How can the Safe System fix a problem like this? It can't, because the Safe System is principally concerned with

---

141   The gap in the centre line removed the one possible visual cue that things are different. Tragically, this gap was not actually required: a driver can legally cross a single solid line (except to overtake) or a dashed line.

reducing kinetic energy to *reduce the consequences of crashes*. But no amount of slowing vehicles to reduce the kinetic energy will fix this problem. A truck driving into the side of a car at 60 km/h will still cause serious injury and a lower limit would not be credible in this environment.

However, the Safety Star System can fix it because the issue is the limit of the human mind to process information and the solution is to *stop the confusing conflicts happening in the first place*. It is evident that unfamiliar drivers are being led into a trap by misleading information: the need to give way is not at all apparent. In other words, the decision is beyond the ability of the human mind to process without prior experience. An understanding of human decision making (part of understanding road user behaviour) allows this assessment to be made. To overcome the fatal and serious injury problem the whole crash problem needs to be tackled. The solution is to align the perception of risk with the actual risk by either restoring normal T-intersection priority (also adding new direction signs) or flattening the corner and creating a new conventional T-intersection for the minor leg.

The Six Star Points

The Safe System has 'three pillars': safer speeds, safer vehicles and safer roads & roadsides (see Figure 9.1). These inadequately describe the factors that road safety needs to consider. The Safety Star System has six principal points:

- ▶ The Road

- ▶ The Road User

- ▶ The Vehicle

- ▶ Laws

134

▶   Speed Management, and

▶   Travel and Planning Policies

All six of these points are areas that need to be considered in a systematic approach to road safety and it is important they are not considered in isolation from each other. Figure 10.1 shows how elements are interconnected. For example, when considering the road user (Star point B), it's not just about enforcement and providing information/ education — it also draws on the additional area of applying an understanding of road user behaviour in order to effectively engineer the road environment to elicit safe road user responses.[142] See all of the supporting stars that have element 'B' listed.

Resources and Inputs

The resources and inputs into these six star points are categorised into 13 distinct areas, which are the supporting stars in Figure 10.1.

The Safe System framework has just four inputs into its three pillars. While this may be simpler to focus on, it is too simplistic and omits some important areas.

The three most important inputs into the six points of the Safety Star System are:

▶   Understanding road user behaviour,

▶   Understanding risk, and

▶   In-depth crash analysis.

---

142   There is also a whole new area of potential road user responses about to open up if automated vehicles take off as some believe they will. How will drivers — and pedestrians — respond to vehicles that have increasing levels of automation, and more 'aids to human capabilities'? Will people become less vigilant, less alert, less caring of consequences because the technology will look after them?

These feed into so many other areas. Without an understanding of road user behaviour we can't design our roads and streets effectively. Road safety engineering knowledge and skills also depend on it. An understanding of risk (i.e. severity *and* likelihood, rather than just severity under the Safe System) allows us to be effective in all six points on the star. In-depth crash analysis has allowed us to understand safety for vehicles (through crash testing) and the relationship between various road arrangements and possible crash types (through blackspot investigation).

No efforts to improve road safety will succeed without the input of adequate resources, both money for infrastructure and engineering improvements (better roads, safer intersections, blackspot fixes, etc.) and also funding of sufficient numbers of skilled and experienced people to design and manage roads. Much of the divergence into ineffective directions in road safety in recent years (including the Safe System) is due to the loss of skills and experience in state road authorities over the past 25 years, as discussed in Chapter 9.

Road safety in town planning has almost been forgotten since the days of the Federal Office of Road Safety's *Planning for Road Safety* report.[143] Or it's been ignored by town planning fads like New Urbanism. The work of Ray Brindle, for example[144] needs to be up-front and central in town planning; the way streets and roads in new urban areas are set out will set down forever their likely levels of road safety.

We need all the inputs and resources shown in the Safety Star System, including the reasonable enforcement of effective laws. But the most effective inputs will be those that provide us with safer infrastructure, because safer infrastructure is not just about designing roads and intersections to various standards — it's about designing with an understanding of road user behaviour. As I have stated elsewhere (Morgan, 2017), to successfully engineer (or re-engineer) the road — or more generally, manage the road — we need to:

143 Department of Transport Office of Road Safety, 1984.
144 Brindle, 1995, Austroads, 2000 and Brindle, 2001.

- ▶ Ask how do we want road users to respond?

- ▶ Understand how they will actually respond.

- ▶ Match treatments with expectations (e.g. make sure roundabouts and other treatments look different from each other).

- ▶ Match treatments with road user capabilities, and

- ▶ Provide consistency.

Roads designed with these considerations in mind will provide a lasting, safe legacy.

Chapter 11

THE LESSONS OF HISTORY

This chapter title was used by Trinca et al, 1988 as a section heading. Concluding their chapter on the role of institutions they firstly list 'the factors which have combined to provide a still sub-optimal traffic safety system in the motorised nations'. Let me paraphrase and summarise their list[145]:

▶ First there was inadequate data, which led to incorrect conceptions of the traffic safety problem.

▶ One wrong conception was that, with most crashes involving 'deficient and/or reprehensible behaviour' by road users, the solutions must lie with behavioural change. This was partly due to the primary road safety agency being the police.

▶ This concept militated against scientific study and the transfer of knowledge from areas like aviation safety.

▶ It also meant road, traffic and vehicle engineering professions were slow to accept accountability, and

▶ There was little coordination between the relevant agencies as the need for an integrated approach was not appreciated.

Trinca et al then describe the key lessons learnt, which are put forward as "five statements of 'attitude' towards traffic safety which we believe

---

145 Trinca et al, 1988, pp 58-59.

need to be adopted by decision-making bodies, both public and private. This attitude set underpins a 'mature' approach to traffic safety."

I believe it is worth quoting these five statements in full:

1. Rationality

Traffic injury is a heterogeneous phenomenon [diverse in character or content] and as such is amenable to rational cause and effect analysis

It comprises [a] myriad definable problems, each arising out of the operation of some aspect of the road transport system. Individual problems can be identified, understood and tackled by the application of scientific techniques, provided adequate data exist.

2. Limited Objectives (a corollary to No. 1)

There can be no traffic safety panacea in a mobile society.

The objective is to bring as many of the defined problems under control as possible, rather than to 'prevent' all crashes.

3. Systems Approach

The component problems of traffic injury are too diverse for any single agency to handle alone.

It is necessary to adopt a system-wide strategy and to achieve program integration through meaningful institutional co-operation.

4.   Cost Effectiveness

Rational decision-making is critical when choosing between competing social and economic objectives which will (and should) always influence traffic safety policy and program development.

The techniques of cost-effectiveness analysis should be applied to assist in conflict resolution based on best available objective evidence.

5.   Pilot Testing and Evaluation

Countermeasure programs, once implemented, take on a life of their own and, if later proven ineffective, are very difficult to remove or modify.

As a result, a proportion of the scarce funds available remain inappropriately applied. Consequently, traffic safety programs should, wherever possible, be pilot tested before large-scale implementation is attempted and evaluation must be an integral part of any program proposal.

How much do current initiatives under the Safe System banner accord with these considered statements? For example, how much scientific rigor is at the forefront these days, rather than dogma? Where is cost-effectiveness in the Safe System? Is not the Safe System with its management of kinetic energy and 'the primacy of speed management' (i.e. low speed limits) offered as the panacea for all road safety problems? Is it honest to call the Safe System a 'system' at all? Is not the Safe System simply the resurgence of efforts at direct behaviour change dressed up as some scientific system? Fatality rates have certainly reduced in the past thirty years, but might it just be possible that without the Safe System, though with the retention of skills and experience in state road authorities, casualty rates as well as fatality rates would be even lower by now?

It concerns me just how far our understanding of road safety has gone backwards since Trinca et al provided their assessment thirty years ago.

In the 1950s and 60s there was a general belief that road users were to blame for crashes and they just needed to be more careful. This allowed governments to maintain disinterest and avoid spending money to address the problem. By 1970 it was out of control and Australia had the highest per capita fatality rate in the world.[146] Nonetheless informed voices outside government were beginning to be heard. The 1968 report by the South Australian Government Committee of Enquiry into Road Safety said:

> Many of the previous publicity campaigns have been directed at drivers to persuade them to change their presumed delinquent behaviour. This absolves from blame, perhaps unfairly, the many organisations and authorities, such as vehicle manufacturers, road authorities and developers whose activities also influence road safety.
>
> In particular, posters with a general message such as 'Speed Kills', 'Drive carefully', 'The Life You Save May Be Your Own' are ineffective.[147]

That was 50 years ago. Figure 11.1 is now.

Thirty years ago Trinca et al offered the following advice to *motorising* countries, not ones as advanced as Australia:

> There is a fundamental need to treat traffic safety scientifically. Risk can be reduced, and deaths, injuries and disabilities diminished through the rational application of resources to effective social

---

146 Quayle, 2015, p 38.
147 Quoted in Quayle, 2015, p 115.

Figure 11.1
Australia 2017, returning to the pre-scientific era of road safety

and technological programs. ... pre-scientific notions concerning simplistic behavioural change inhibit progress.[148]

It's not some amazing coincidence that we see things like those in Figure 11.1 at the same time as the Safe System is the cornerstone of all road safety strategies. The underlying causes are the same: the tragic loss of skills and experience in state road authorities, allowing the return to pre-scientific notions of road safety 'solutions'. Only this time there is the potential for unlimited automated enforcement to cover over the ineffectiveness. As Quayle comments[149], in the 1950s

---

148 Trinca et al, 1988, p 76.

149 Quayle, 2015, p 301.

it was 'If only people would be more careful'. Now it's 'If only people would slow down'.

Imagine what aviation safety would look like if it adopted road safety's Safe System approach. Instead of investigating each plane crash, we'd blame or lock up the pilot. Instead of detailed investigation of whether any systems, rules, mechanical faults or human errors contributed to a crash or incident, we'd tell pilots to slow down and travel at a safe flying speed. We'd get aircraft designers to make cockpit windows bigger so pilots could be more careful and keep a proper lookout. We'd put big buffers in place in case they ran off the runway, instead of trying to prevent crashes in the first place. We certainly wouldn't investigate the possible causes of any human error and change systems to prevent those errors happening and leading to crashes. We'd say our vision is zero deaths and serious injuries and spend a lot of time telling everyone about our vision and how they all need to play their part because it's a joint responsibility. Amusing perhaps, but there is also a sad irony in this. Much of what was learnt in Australia 50 years ago about human factors in road safety — and which now effectively plays no part in the Safe System — came from scientific study by Ron Cumming (see Figure 3.1) and other people who transferred across from aviation safety.

This chapter started with five dot points summarising why early efforts in road safety had 'sub-optimal' results. Reflect on those while you look at Figure 11.2, a photograph taken at the launch of Victoria's *Towards Zero* road safety initiative in 2015.

Accompanying the minister at the launch were representatives of just two government organisations: Victoria Police and the state's no-fault insurer, the TAC. One has a focus on law enforcement; the other has a direct financial interest in focusing on fatal and serious injury reduction. What are we to make of the Zero the CEO of the TAC is holding? Is it the ultimate vision of no deaths and serious injuries on our roads? If so the TAC (and anyone else who believes this a worthy objective) needs to explain not how we get from where we are now to that zero position, but rather what that zero position actually looks like. What does it look

Figure 11.2
The Launch of Victoria's *Towards Zero* initiative on 26th August 2015:
Road Safety Minister Luke Donnellan (centre) accompanied by Mr Joe
Calafiore, CEO of the Transport Accident Commission and Assistant
Commissioner Doug Fryer, Victoria Police (Photo courtesy of TAC).

like in terms of roads and streets? More importantly, what does it look like in terms of other community desires and interests like taxation, government finances, police activity, the 'justice' system, freedom of speech, freedom of activity (do we ban other risky activities as well?), air pollution, global warming, mobility, equality of opportunity, job security, etc.. Because if zero is not negotiable and it's road safety at any cost, what is the cost of that end point? The consequences of that end point need to be described in some detail.

For me the Zero represents 'absent friends': all the skilled and experienced road safety engineers, road designers and other engineering and scientific professionals who could provide Victoria (and Australia) with real and effective road safety improvements, but who are no longer employed by the state road authorities.

Taking a broader view than just road safety, the Zero might represent all the other aspects of society that influence road user behaviour and levels of road safety, but are not acknowledged in a *system* that believes road safety is paramount, the utopia of total safety is attainable and its achievement will come about through adhering to Figure 9.1.

What of the future?

How will the Safe System, Vision Zero, Towards Zero and their ilk be viewed a decade from now? What about two decades from now? When fatalities and serious injuries from road crashes have been significantly reduced — say by 30 to 50% — but are still nowhere near zero? Or will it be a case of early gains and then a plateauing out? What then? Will we see ever more control over our lives in the name of 'life is sacrosanct' or 'if we only save one life it's worthwhile', with little trauma reduction to show for it? Or will governments and road authorities realise their promise of zero leaves them more liable the longer there's no prospect of it and reframe zero into something even less than *towards* zero?

Let's not wait to find out. It's time to dump the Safe System and in its place:

▶ Adopt the Safety Star System in Figure 10.1 and the philosophy behind it.

▶ Dramatically increase the number of skilled and experienced road safety engineers (and road designers and other engineering and scientific professionals) employed directly in state and territory road authorities.

▶ Put an emphasis on long-lasting, cost-effective infrastructure safety improvements on our road system.

▶ Stop the 1960s approach of trying to directly modify road user behaviour. Accept the reality behind Figure 2.1.

▶ Return to a credible speed management system, similar to Table 7.1 and acknowledging the points in the second half of Chapter 9.

▶ Restore the place of genuine, independent road safety research. Restore the principle of cause and effect. Restore the requirement to establish a safety *problem* before offering a safety *solution*. Restore the application of crash and blackspot investigation.

▶ Focus enforcement on the *deviant* or *high-risk* 10% (or 4%) and not on the majority.

▶ Be honest that there is a limit to what we can achieve in bringing safety to an inherently risky activity.

Zero is not possible; attempts to achieve it will have severe adverse consequences, including an invasive control over many aspects of our lives. And for what? It is clear from the references in Chapters 5 and 7 and the speed summary at the end of Chapter 9 that seeking to make us all travel at unreasonably low speeds will not bring about the claimed reductions in casualty crashes. Focusing on fatal and serious injury crashes will result in fewer fatal and serious injury crash problems being fixed, than if we address all crash problems.

Why have I, in Chapter 1, called the Safe System ideas, proposals and recommendations from Johnston et al, 2017 a *Manifesto*? Because it is political. It takes elements of science and mixes them with dogma. It claims life is sacrosanct but cares little about freedom. Mobility is unimportant. It discredits 85th percentile speeds — the basis of effective speed limits. It virtually ignores the role of infrastructure in the history of accident trauma reduction.[150] It impinges on all of us, yet addresses just one percent of all crashes.

---

150 As noted in Quayle, 2015 at p 305: leaving aside random breath testing and vehicle improvements (occupant restraints, head protection, better crashworthiness and interior design), the reduction in the risk of a crash has been achieved most notably by the duplication of major highways, the building of urban freeways, local area traffic management measures, traffic control at intersections and high standard street lighting.

Table 11.1

A comparison of the Communist Manifesto
and the *Safe System Manifesto*

Aspect of Each Manifesto	The Communist Manifesto	The Safe System Manifesto
Objective: the utopia of …	Zero oppression ('Freedom from oppression').	Zero fatal and serious injury crashes.
Means to achieve it:	Communism.	The Safe System.
Claimed to be a science?	Yes, by Engels.	Yes, by Mooren et al, 2011.
Heralded as a new and ultimate concept on the topic?	Yes, 'The end of history' when freedom is achieved.	Yes, 'A breakthrough in road safety thinking'.
But actually suited to …	19th century politics.	Current times with the loss of road authority skills and experience.
Demands the destruction of …	Capitalism.	The existing speed limit regime based on 85th percentile speeds.
Compromise allowed?	No — other points of view repressed.	No — road safety is everything. Zero is not negotiable.
Requires enforcement?	Yes — extensive secret police network: unseen.	Yes — extensive speed camera network: automated.
Actually results in …	An unchecked state. Oppression, starvation, mass extermination. A police state. Control of the population.	An unchecked road safety juggernaut. Ongoing fatal and serious injury crashes. A police state. Control of the population.

In Table 11.1 I compare the *Safe System Manifesto* with Marx and Engels' Communist Manifesto which led to the Stalin regime in the Soviet Union.

Such noble intentions and such unfortunate, unintended consequences.

The history of humanity is full of episodes of great expansions, brilliant insights and brave adventures. All involved risk. But history is also littered with episodes of retreat, stagnation, forgotten knowledge and demise.

Which will the Safe System be viewed as? A brilliant breakthrough in thought that led us to yet more breakthroughs and greater well-being — or an idea that risked leading us to being controlled, cautious and more anxious, forgetting past knowledge and being guided by dogma? To paraphrase Kotkin, 2017, the Safe System's demand to abolish the old order of evidence-based road safety and speed management has been a clarion call to action and — unless we put a stop to it — its continuation will put us on an inexorable path to the creation of an unchecked state.

The road to ruin is paved with good intentions and increasingly these days it has a low speed limit and frequent speed cameras. The epitaph on the headstone of the Safe System will simply say:

THEY MEANT WELL

Acknowledgements

If there is one thing I've learnt through my career it is that there is much yet to learn. The knowledge I have gained while writing this book has been significant. The extensive list of references just hints at the work that has gone on before today in endeavours to improve safety on our roads. So thank you to those whose work is referenced here.

Thank you to those who have given their permission for the reproduction of many of the images used in this book (all credited adjacent to the image).

To everyone who has provided comment, feedback and advice on drafts of this book and ideas I have developed in it, I would like to say thank you. This includes Ray Brindle, Ken Ogden and Judd Epstein. Some of that advice I actually took notice of.

Thanks to Anthea Wynn for applying her editing knowledge and skills and to Luke Harris of WorkingType Studio for the great cover design and all his production skills, without which this book would not have eventuated.

And thanks to Lucy for her support while I was writing this book.

References

Adams, J 1985, *Risk and Freedom,* Transport Publishing Projects, Cardiff.

Andreassen, D *2013, Submission to the Inquiry into Serious Injury,* Parliament of Victoria Road Safety Committee, Melbourne.

ATSB: Australian Transport Safety Bureau 2003, *International Road Safety Comparisons, The 2001 Report,* Australian Government, Canberra.

Australian Transport Council 2011, *National Road Safety Strategy 2011-2020,* Canberra.

Austroads 1994, *Road Safety Audit,* Report AP-30/94, Austroads, Sydney.

Austroads 2000, *A Framework for Arterial Road Access Management,* Research Report AP-R163, Austroads, Sydney.

Austroads 2004, *Guide to Traffic Engineering Practice, Part 4 Treatment of Crash Locations,* Austroads, Sydney.

Austroads 2008, *Guide to Road Safety, Part 3 Speed Limits and Speed Management,* AGRS03-2008, Austroads, Sydney.

Belin, M, Johansson, R, Lindberg, J & Tingvall, C 1997, *The Vision Zero and its Consequences,* 4th International Conference on Safety and the Environment in the 21st Century, November 23-27, Tel Aviv, Israel.

Bergh, T, Remgård, M, Carlsson, A, Olstam, J & Strömgren, P 2016, *2+1-roads Recent Swedish Capacity and Level-of-Service Experience,* Transportation Research Procedia Volume 15, 2016, pp 331–345, Elsevier, Amsterdam.

BITRE: Bureau of Infrastructure, Transport and Regional Economics 2009, *Road crash costs in Australia 2006,* Report 118, Canberra.

BITRE: Bureau of Infrastructure, Transport and Regional Economics 2017, *International road safety comparisons 2015,* Statistical Report, Canberra.

Bobevski, I, Hosking, S, Oxley, P & Cameron, M 2007, *Generalised Linear Modelling of Crashes and Injury Severity in the Context of the Speed-Related Initiatives in Victoria during 2000-2002,* Monash University Accident Research Centre, Clayton.

Brindle, RE 1995, *Living with Traffic: Twenty-seven contributions to the art and practice of traffic calming 1979 – 1992,* ARRB Transport Research Ltd Special Report No. 53, Vermont South.

Brindle, RE 2001, *Planning and road safety: Opportunities and barriers,* paper presented at Australasian Transport Research Forum, Hobart.

Campbell, M 2016, *Speed cameras go covert,* Drive (Fairfax Media) 3rd October, Sydney.

Carlsson, A 2009, *Evaluation of 2+1 roads with cable barrier,* Final Report, VTI report 636A, Linköping, Sweden.

Clark, N & Pretty, RL eds. 1969, *Traffic Engineering Practice,* 2nd edn, University of Melbourne, Melbourne.

Corben, B, Logan, DB, Johnston, I & Vulcan, P 2008, *Development of a new road safety strategy for Western Australia 2008-2020,* Monash University Accident Research Centre Report No. 282, MUARC, Clayton.

Crinion, JD 1969, *The Effects of Traffic Engineering Measures* in Clark N & Pretty RL eds, *Traffic Engineering Practice*, 2nd edn, University of Melbourne, Parkville.

Cumming, RW 1964, *The analysis of skills in driving*, Australian Road Research 1(9), pp 4-14, Australian Road Research Board, Vermont South.

Cumming, RW & Cameron, C 1969, *Driver Response to Traffic*, in Clark N & Pretty RL eds, *Traffic Engineering Practice*, 2nd edn, University of Melbourne, Parkville.

Department of Transport Office of Road Safety 1984, *Planning for Road Safety*, AGPS, Canberra.

Ekman, L 2014, pdf of slide presentation: *http://www.bbars.bg/storage/ pdf/kragla masa 03.10.15/inj.P.Tabakov – 2+1-Roads-Sweden.pdf.*

Goodyear, S 2014, *The Swedish Approach to Road Safety: The Accident Is Not the Major Problem,* Interview with Dr Matts-Åke Belin, from Trafikverket, Swedish Transport Administration, Citylab.com, 20 November, Atlantic Media, Washington.

Grzebieta, R & Rechnitzer, G 2001, *Crashworthy Systems – a paradigm shift in road safety design (part II)* Transport Engineering in Australia, IEAust, Vol. 7, Nos. 1&2, Sydney.

Haight, FA 1994, *Problems in Estimating Comparative Costs of Safety and Mobility,* Journal of Transport Economics and Policy, Vol. 28, No. 1, January, Bath, UK.

Harper, S 2009, *Essay Review: Rose's Strategy of Preventive Medicine. Geoffrey Rose with commentary by Kay-Tee Khaw and Michael Marmot*, International Journal of Epidemiology, Vol. 38, issue 6, Dec., OUP, Oxford, UK.

Highly, K 2016, *Dancing in My Dreams, Confronting the Spectre of Polio*, Monash University Publishing, Clayton.

Hillier, P, Makwasha, T & Turner, B 2016, *Achieving Safe System Speeds on Urban Arterial Roads: Compendium of Good Practice,* Austroads Research Report AP-R514-16, Sydney.

ITF 2017, *Road Safety Annual Report 2017,* OECD Publishing, Paris. *http://dx.doi.org/10.1787/irtad-2017-en.*

Johnston, IR, Muir, C & Howard, EW 2017, *Eliminating Serious Injury and Death from Road Transport, A Crisis in Complacency,* CRC Press, Boca Raton, FL, USA. Originally published in hardback in 2014.

Jurewicz, C, Tofler, S & Makwasha, T 2015, *Improving the Performance of Safe System Infrastructure: Final Report,* Austroads Research Report AP-R498-15, Sydney.

Jurewicz, C, Sobhani, A, Chau, P & Woolley, J 2017, *Understanding and Improving Safe System Intersection Performance,* Austroads Research Report AP-R556-17, Sydney.

Kloeden, CN, McLean, AJ, Moore, VM & Ponte, G 1997, *Travelling Speed and the Risk of Crash Involvement,* Report CR 172, Federal Office of Road Safety, Canberra.

Kotkin, S 2017, *A Century of Bloodshed,* Inquirer in The Australian, p 11, 6 Nov, Sydney.

Kruger, J & Dunning, D 1999, *Unskilled and Unaware of it: How Difficulties in Recognizing One's Own Incompetence Lead to Inflated Self-Assessments,* Journal of Personality and Social Psychology, Vol 77, No. 6, pp 1121-1134, Washington.

Lambert, J 2002, *The Magical Property of 60 km/h as a Speed Limit?* Road Safety 2002 Conference, ACRS, Canberra.

Langford, J & Oxley, J 2006, *Assessing and managing older drivers' crash risk using safe system principles,* 7th International Conference on Walking and Liveable Communities, Melbourne.

Leeming, JJ 1969, *Road Accidents Prevent or Punish?* Cassell & Company, London.

Leeming, JJ 1977, *Road Accidents: The Armoured Door*, Faculty of Building, Borehamwood, UK.

Lydon, M & Turner, B 2017, *Building a Safe System for Transport* in Delbosc, A & Young, W eds, *Traffic Engineering and Management*, 7th edn, Monash Institute of Transport Studies, Clayton.

McInerney, R 2002, *The Road Safety Risk Manager: a software tool to help practitioners prioritise, manage and track road safety engineering issues,* Proceedings of the Road Safety Research, Policing and Education Conference, pp 189-193, Vol 2, Adelaide.

Main Roads Western Australia 2012, Main Roads WA Road Safety Strategy 2011-2015, *The Road Towards Zero – No more death or serious injury on our roads,* MRWA, Perth.

Meesmann, U & Rossi, M, 2015, *Drinking and driving: learning from good practices abroad*, Belgian Road Safety Institute – Knowledge Centre Road Safety, Brussels, Belgium.

Miller, N 2018, *Hypocrisy? What's the problem?* Last Word column, The Age, 13 January, Melbourne.

Mooren, L, Grzebieta, R & Job, S 2011, *Safe System – Comparisons of this Approach in Australia*, Australasian College of Road Safety Conference 'A Safe System: Making it Happen!' 1-2 September, Melbourne.

Morgan, R 1988, *Left Turn Versus Right Turn Priorities – What Can Australia Learn from New Zealand?* Australian Road Research 18(1), March, pp 11-20, Australian Road Research Board, Vermont South.

Morgan, R 1994, *Letter to the Editor,* Institute of Transportation of Engineers Journal, 14 Feb, Washington.

Morgan, R 2007, *S is for Sweden, Speed Limits and Skills Shortage,* AITPM Newsletter September/October, Australian Institute of Traffic Planning and Management, Toombul, Qld.

Morgan, R 2014, *What Went Wrong With Road Safety Auditing?* AITPM National Conference, Adelaide.

Morgan, R 2017, *Myth-Placed Traffic Engineering,* AITPM Victorian Branch technical forum 21st September *(https://www.aitpm.com. au/wp-content/uploads/2017/04/Myth-Placed-Traffic-Engineering-Robert-Morgan.pdf),* AITPM, Toombul, Qld.

Moses, P 1989, *Young and Old – A Study in Pedestrian Safety,* Australian Transport Research Forum papers, DIRDC, Canberra.

Moskvitch, K 2013, *Penal Code,* New Scientist, 7 Sept., pp 37-39, London.

Newstead, SV & Mullan, NG 1996, *Evaluation of the crash effects of the changes in speed zones in Victoria during 1993-1994 (excluding 100 to 110 km/h),* Monash University Accident Research Centre, Report No. 98, MUARC, Clayton.

OECD/ITF 2008, *Towards Zero, Ambitious Road Safety Targets and the Safe System Approach,* OECD Publishing, Paris.

Office of Road Safety, Western Australia 2009, *Towards Zero – Road Safety Strategy,* ORS, Perth.

Ogden, KW 1996, *Safer Roads: A Guide to Road Safety Engineering,* Avebury Technical, Aldershot, UK.

Proctor, S 2017, *Safe Systems and Road Safety Audit,* TMS Consultancy presentation to Scottish Road Safety Conference 'Safer Roads for Everyone', 3rd April, Stirling, UK.

Quayle, G 1999, *Hit by Friendly Fire – Collateral Damage in the War Against Speed,* ACRS National Conference, Canberra.

Quayle, G 2015, *Driving Past – A Memoir of What Made Australia's Roads Safer*, Balboa Press, Bloomington, IN, USA.

Rose, G 1981, *Strategy of prevention: Lessons from cardiovascular disease*, British Medical Journal 282, pp 1847-1851, London.

Ross, HL, Klette, H & McCleary, R 1992, *Recent Trends in Scandinavian Drunk Driving Law*, 9th International Conference on Alcohol, Drugs and Traffic Safety, September, Cologne, Germany.

Silcock, D, Smith, K, Knox, D & Beuret K 2000, *What Limits Speed? Factors that affect how fast we drive*, AA Foundation for Road Safety Research, UK.

Smart, W, de Roos, M, Job, S, Levett, S, Tang, J, Graham, A, Gilbert, L, Hendry, T, Foster, J & O'Mara W 2009, *The Newell Highway Road Safety Review*, 2009 Australasian Road Safety Research, Policing and Education Conference, Sydney.

Standards Australia 1986, AS 1742.4 *Manual of uniform traffic control devices, Part 4 – Speed Controls*, Sydney.

Standards Australia 1999, AS 1742.4 *Manual of uniform traffic control devices, Part 4 – Speed Controls*, Sydney.

Standards Australia 2008, AS 1742.4 *Manual of uniform traffic control devices, Part 4 – Speed Controls*, Sydney.

Stedman Jones, G 2016, *Karl Marx Greatness and Illusion*, Penguin, London.

Sunshine Coast Council 2016, *Sunshine Coast Council's Road Safety Plan 2016-2020*, Maroochydore.

SWOV 2013, Fact sheet, *Sustainable Safety: principles, misconceptions, and relations with other visions*, SWOV, Leidschendam, Netherlands.

Trinca, GW, Johnston, IR, Campbell, BJ, Haight, FA, Knight, PR, Mackay, GM, McLean, AJ & Petrucelli, E 1988, *Reducing Traffic Injury – A Global Challenge*, Royal Australasian College of Surgeons, Melbourne.

Transport for NSW 2012, *NSW Road Safety Strategy 2012-2021*, TfNSW, Chippendale.

Truong, J & Cockfield, S 2015, *Towards Zero – Building a safe road system for Victoria*, Proc 2015 Australasian Road Safety Conference, 14-16 Oct, Gold Coast, Queensland.

Turner, B & Jurewicz, C 2016, *Development and use of the Austroads Safe System Assessment Framework*, Proc 2016 Australasian Road Safety Conference, 6-8 Sep, Canberra, ACT.

Turner, B & Lydon, M 2017, *Safe System Solutions* in Delbosc, A & Young, W eds, *Traffic Engineering and Management*, 7th edn, Monash Institute of Transport Studies, Clayton.

Turner, B, Lydon, M, Reynolds, J & Sobhani, A 2017, *Understanding road safety problems* in Delbosc, A & Young, W eds, *Traffic Engineering and Management*, 7th edn, Monash Institute of Transport Studies, Clayton.

Turner, B, Partridge, R, Turner, S, Corben, B, Woolley, J, Stokes, C, Oxley, J, Stephan, K, Steinmetz, L & Chau P 2017, *Safe System Infrastructure on Mixed Use Arterials*, Austroads Technical Report AP-T330-17, Sydney.

About the Author

Rob Morgan is based in Melbourne, Australia. He is one of the country's most experienced road safety engineers, with over four decades in road safety engineering, traffic engineering and traffic planning across Australia and overseas. He has been the principal author of several national road safety engineering guidelines and a lead presenter at training workshops on these topics in most Australian jurisdictions. From 1991 to 1993 he was a member of VicRoads' Speed Management Policy Committee and from 2000 to 2009 he was a member of the national Road Rules Maintenance Committee. Rob is also a long-time member of the committee responsible for Australia's Manual of uniform traffic control devices; major aspects of our standard parking signs, direction signs, street name signs and freeway signs are the result of his initiatives.

Author photograph by Lucy Colangelo.

www.ingramcontent.com/pod-product-compliance
Lightning Source LLC
Chambersburg PA
CBHW070727220326
41598CB00024BA/3336